Microcircuits
of
Capital

Human Geography

Series Editor: Derek Gregory

Published

Nigel Thrift, *Social Theory and Human Geography*
Michael Dear and Jennifer Wolch, *Landscapes of Despair*
John Eyles and David M. Smith, *Qualitative Methods in Human Geography*
Kevin Morgan and Andrew Sayer, *Microcircuits of Capital*

Forthcoming Titles

Derek Gregory, *An Introduction to Human Geography*
Stephen Daniels, *Landscape, Image, Text*
Susan J. Smith, *White Supremacy in Britain?*

Microcircuits
of
Capital

'Sunrise' Industry and Uneven Development

Kevin Morgan and Andrew Sayer

Polity Press

Copyright © Kevin Morgan and Andrew Sayer 1988

First published 1988 by Polity Press
in association with Basil Blackwell

Editorial Office:
Polity Press, Dales Brewery, Gwydir Street,
Cambridge CB1 2LJ, UK

Basil Blackwell Ltd
108 Cowley Road, Oxford OX4 1JF, UK

British Library Cataloguing in Publication Data

Morgan, Kevin
Microcircuits of capital: 'sunrise'
industry and uneven development.
1. Electronics industries
I. Title   II. Sayer, Andrew
338.4'7621381

ISBN 0-7456-0339-4

Typeset in 10½ on 12pt Sabon
by Gecko Limited, Bicester, Oxon.
Printed in Great Britain by
T. J. Press (Padstow), Cornwall

For Elizabeth and Louis

Contents

List of Figures and Tables

Figures

Tables

Preface

This book had its origins in two ESRC-funded research projects: the first project began in 1982 and looked at the electrical engineering industry in South Wales (project number HR 8353(D)); the second, looking at the development of the electronics industry in the so-called M4 corridor, began in 1984 and finished in 1985 (project number DOO 232077). We are grateful to the ESRC for its material support for these projects. The book was conceived and written at a time when enormous claims were (and still are) being made on behalf of 'high technology'. In focusing on the electronics industry our central aim has been to introduce a critical perspective against which these wider claims can be assessed. This is an important political task because 'high technology' is widely perceived to be a panacea for industrial decline, regional decay and unemployment – some of the pressing political issues in contemporary Britain.

All books carry debts, social as well as intellectual, and ours is no exception. The University of Sussex provided a highly favourable environment for interdisciplinary research and here our thanks go to Tony Fielding, James Barlow and Mike Savage in Urban and Regional Studies. Our association with the Science Policy Research Unit at Sussex has been enormously helpful; our thanks especially to Erik Arnold, Tim Brady, Ken Guy and Peter Senker. Thanks also to Susan Rowland for drawing the diagrams. Further afield we must thank Nia and Phil Cooke and Eirlys and Kim Howells and Terry and Gareth Rees for their 'hotels'; Gary Lawson of the Engineering Industry Training Board; Martin Boddy and John Lovering of the School of Advanced Urban Studies, Bristol; Richard Meegan and CES Ltd; Steve Kennedy, Doreen Massey, Dick Walker, Michael Storper and Dieter Ernst; and finally the many managers, trade union officials and workers who gave so much of their time, without which this book could not have been written. Our main debts are not so far afield, and our thanks and love go to Sue Taylor and Hazel Ellerby for helping us keep things in perspective and for reminding us of the problems of other kinds of 'uneven development'.

PART I

Introduction and Theory

1

Introduction

An American multinational, X, headquartered in Silicon Valley, announces the setting up of a new electronics plant in a British development area, thereby extending its global production system. The news is greeted by the regional development agency as a major boost to the local economy – perhaps the beginnings of another Silicon Valley – and as a sign that it need no longer be dependent on 'smokestack industries' like coal and steel. However the work to be carried out at the new plant and the type of workers are very different from those of X's Silicon Valley plants, and indeed from those of its West German and Taiwanese plants. Central government welcomes not only the promised job creation but the boost to the 'British' electronics industry and to what it sees as the key sector for economic regeneration in the coming decades. It also hopes that indigenous firms will learn something from X's sophisticated management techniques. Other interests see X's arrival differently. To Y, an established indigenous company, it is a competitive threat, and one that has been increased by the government's action in welcoming it into the British market. Trade unions already involved in Y share these doubts but naturally want to ensure that X's plant is unionized. This proves to be difficult and in any case they have mixed feelings about the implications of its sophisticated management techniques which challenge traditional ways of working. A few years later X itself comes under threat from imports from Japan and the newly industrializing countries and announces redundancies in response to a crisis of overproduction in the industry. Meanwhile, Y has already substantially withdrawn from competition in this market, again with resulting job losses in various regions. At national and EEC levels, governments try to rally firms in similar situations to act together against overseas competition although, as many of the firms themselves are in competition with one another and have overseas interests, little comes of it.

To regular readers of the business press there are some familiar themes here, but it is a highly complex story, involving a range of different social, economic and political interests, partly conflicting, partly parallel, operating at scales ranging from the local to the international, and in places as different as California, central Scotland, Tokyo and Penang. It is also poorly understood. Part of the reason for this is the tendency for 'hype' to drive out serious analysis where industries considered to be 'high tech' are concerned. But the main reason is the difficulty of grasping how all these cross-cutting interests arise as part of a process of *uneven development* – a process in which nothing less than the rise and fall of industrial countries and regions is bound up.

In this book we examine that process by showing how these themes interweave in the electronics industry and how they produce very different kinds of development in different places. It is a study of the interrelationships between economic, social, political and spatial dimensions of the development of the industry, concentrating on Britain, but as part of an international division of labour. Within this broad ambit, we pay special attention first to the changing global division of labour and pattern of comparative advantage, looking at the shifting relative positions of the USA, Europe, Japan, the newly industrializing countries and other Third World countries. Secondly, and with particular reference to Britain, we look at the industry's implications for regional development and for labour in terms of employment and the changing nature of work. As our opening hypothetical example suggested, new industries are frequently associated with new forms of social organization of production and in part their success depends on winning assent to them, often in areas with contrasting labour traditions. Finally, in the light of these assessments, we consider the condition of electronics in Britain and its policy implications. Note, however, that we do not deal with the wider question of the economic impact on user industries and the economy at large, but focus instead on processes either internal or immediately contextual to the electronics sector.[1]

Why electronics?

Our approach could be used to study any industry but we have chosen electronics because of its current strategic importance for modern economies. Microelectronics is set to be the 'heartland technology' of economic growth for the next few decades. While some fear that massive technological unemployment will result, others believe that electronics could provide the propulsive force behind a new 'growth ensemble', pulling economies out of the recession, much as the automobile and consumer durable industries did in the 1930s. Countries which lack a strong presence in the industry and are technologically backward are liable to be disadvantaged across a wide range of other user sectors, including arms production, and to become dependent on 'leading-edge' foreign companies. Not surprisingly, virtually every advanced industrial country has a government microelectronics programme and the sector is the prime target of regional and local development agencies.[2]

Moreover, particularly through the technological marvels of 'the chip', electronics has become the focus of great popular interest too; indeed the chip is to the current era what the steam engine was to the mid-nineteenth century. This helps to bolster the enthusiasm of state agencies for promoting the industry, generating interest in 'science parks', 'high tech' and the possibility of new 'Silicon Valleys'. Generally it supports a belief in the industry's possible role as economic saviour of the world's depressed economies. And on the face of it, some of the evidence points this way too, particularly the astonishing dynamism of technical change itself, the exceptional rapidity of growth of many companies and local industrial agglomerations, the role of the industry in the rise of newly industrializing countries like South Korea, and the wealth generated by the industry in places like southern California. But while the Right may cite these aspects there is another side to the industry which tends to be stressed by the Left – the numerical dominance of low paid workers, the tendency for technological change within the industry to displace such people, and the often appalling working conditions they endure in Third World factories. That such polarized views can coexist is itself evidence of uneven development.

It is our belief that the undoubted strategic importance of electronics requires a more sober analysis, particularly because of the dilemmas and contradictions posed by the complex network of cross-cutting interests involved. We have to get beyond some of the stereotyped views regarding uneven development which have become popular in the academic literature, such as the product cycle model (Vernon, 1966) and the New International Division of Labour thesis (Fröbel et al., 1979). To do this we shall have to look at the structure and dynamics of the industry in more detail than is customary in academic writing and avoid selecting only cases which fit such stereotypes. We shall try to establish how far the contrasting aspects referred to above are interdependent and ask what kinds of development are occurring and for whom, and which interests and which areas are likely to gain or lose. It is in the context of answers to such questions that we can consider possibilities for action.

In order to answer these questions we shall, in the course of the book, explore a number of themes, some largely substantive, others involving commentaries on theories.

The international character of the industry

Much of the electronics industry is highly internationalized. As one can easily discover by taking the back off a microcomputer and looking at the names on the chips inside, it is not just that a product bought in one country may bear the name of a foreign company, but that increasingly the production of its component parts is scattered across the world. This fact has a crucial bearing on questions regarding company and national economic performance, regional development and the situation of labour. There is not only an international market for electronics goods but a dense mesh of international production systems belonging to major multinational firms like IBM and Philips. Such firms have affected the differentiation of the world economy by allocating different activities to different areas. At the same time they have also integrated the world economy more tightly, making the activities of workers in different countries more interdependent, both in terms of their roles within an international corporate division of labour and their increasing involvement in international inter-company competition. Even the traditionally protected national markets for telecommunications and defence are opening up to foreign invasions. As we shall see, just how this process works is strongly influenced by the government policies and character of the home bases of the multinationals.

So one can see the same company signs on factories and offices across the world, and yet the activities being carried out behind them may be quite different in each case. Apple in Ireland has a quite different significance from Apple in California, in terms of regional and national impact, kinds of workforce and job security. Regional development agencies tend to play down such differences and boast of their 'catches', parading 'their' company names in the hope of drawing optimistic parallels with major electronics agglomerations like Silicon Valley so as to attract more inward investors. To avoid such illusions, any questions concerning either regional industrial development or the standing of the national industry must be looked at in the appropriate international context. This involves firstly acknowledging the internal functional differentiation of the industry and its firms and secondly taking into account the relative competitive strength of companies and of national electronics industries.

Technology and the economic behaviour of firms

Central to the understanding of any industry is the relationship between its technology – both of the products and of the ways of making them – and its economic characteristics. This is a two-way relationship: the form of the technology and the trajectory of its development both shape and are shaped by the economic circumstances of the industry. This is evident not only at the heart of production in the labour process but in the general organization of sectors and the strategies and problems of individual firms. Consequently little sense can be made of industrial development without some understanding of its technologies, and even within electronics there are major contrasts, e.g. between telecommunications and semiconductors. Moreover, in an industry with such exceptional rates of technological change, it is vital to note the global patterns of technological leads and lags and their associated relationships of technological leadership and dependence, for these structure competition and constrain industrial policy.

The social organization of the industry

Industry is not simply a technical or purely economic field; it also has a social dimension, not just in its external social impact, but in its internal organization: a firm is a set of social relations. Again their character both shapes and is shaped by the immediate technology and economic circumstances of the firm and the wider socio-economic and cultural milieux. National and sometimes regional differences in these milieux are therefore apparent in the character of social relations in firms from different places. They are particularly striking in the contrasts between American, British and Japanese firms and they can make a considerable difference to competitive performance.

In the history of capitalism new firms and industries have frequently been innovative in their social, as well as their technical, organization, so as to facilitate the realization of their technical potential. Multinationals play a major role in diffusing new practices; where one sets up in a country with established practices different from its own, as in the case of firm X in our opening example, it is liable to be welcomed and imitated in some quarters and resisted in others. While the result may be compromise, a number of such entries into a country may significantly change the nature of its work and work culture. Given that electronics is one of the few sectors experiencing rapid growth and frequent inward investment and given the effects of the recession in reducing labour resistance, it turns out that it is playing a leading role in establishing new management – labour relations and work practices. Management – a subject neglected by both neoclassical and radical economics – is therefore on our agenda and we shall be examining the process of social innovation more closely in two radically contrasting regions of Britain.

Regional development

Much has been written on the regional impact of new industries. In recent years the role of technological change in regional development has become a popular topic. The reasons for this are many, but at base they reflect a recognition of the propulsive effect of technological change and the association of successful industrial agglomerations with innovation. However, the evidence behind such beliefs is again highly ambiguous. We shall examine the effect of the industry on regional development, regarding firstly how firms create a 'spatial division of labour' between different regions, secondly how they manage to perform in such contexts, thirdly what their regional impact is in terms of local purchasing and subcontracting and fourthly who actually benefits in the localities in which electronics firms set up. Combining these last two points, we shall ask whether regional development amounts to development *of* a region or merely development *in* a region, i.e. the 'cathedrals in the desert' syndrome.

The development and condition of the British electronics industry

There have been marked disagreements about the industry's state of health in Britain. While the Thatcher Government regards it as a flourishing industry the National Economic Development Council sees it as gripped by relative decline. Here, much depends on the criteria of success and the definition of 'British electronics', but note too that both the problem (if it is one) and the definitional ambiguity are again aspects of uneven development. In any case, success in capitalism is always relative – in this case relative to the performance of world-leaders in the USA and Japan. Again, as with assessments of regional development we must ask, development for whom? Not only this, it is pertinent to ask, development of what? – for electronics firms can be making anything from missile guidance systems to video recorders or medical equipment.

Government–industry relations and the political dimension of competition

Here we are concerned with the relationship between the industry and the state at national and local levels and political processes within the industry itself. Partly for the reasons already given, governments have played an important role in the industry, influencing the lines of competition and mediating the relationship between capital and labour. Often, the industry itself has actively solicited this intervention; indeed the history of the industry has been characterized by corporate politicking both inside and outside the state. As we shall see, this political dimension has been an integral part of the competitive process in the industry rather than an occasional and exceptional intrusion. Nevertheless, there is a considerable variation in the forms of government–industry relations in different countries and we shall pay special attention to this diversity, both as it affects the course of development and as a source of alternative strategies for the future.

The ideology of high tech

In response to the tendency for 'hype' to substitute for analysis where high-technology industry is concerned, we shall repeatedly compare the reality of the electronics industry and its impact with the popular image and the expectations attached to it. Although this ideology presents a distorted picture of certain industries it cannot be ignored since it affects actual developments, particularly ones involving state agencies; however misleading, it is part of what has to be explained.

Although this combination of interests is unusual, all these themes interact closely and are indispensable for explaining actual instances of development in the industry; an interdisciplinary approach is essential.

As regards theoretical orientations, our approach was both attracted and repelled by a number of bodies of literature. One of the main themes of the book is the *spatial* uneven development of the industry. To some readers this may seem strange for in much of social science there has long been a tendency to separate process from space, ignoring the latter as if it made little or no difference or as if spatial patterns were only of incidental interest as a by-product of development (Massey, 1984; Gregory and Urry, 1985). Yet spatial arrangements, at whatever scale, make a difference to processes such as industrial development and to people's lives. It matters to buyers of computer systems that service engineers are close at hand, that firms have good access to specialized local labour markets; it matters whether indigenous firms are protected from foreign competition by distance and that areas of cheap labour may be distant from the richest markets and from R&D centres, and so on. Improvements in transport and communications technology may have reduced the barriers of distance but in many activities face-to-face contact is still felt to be necessary. Such developments increase, rather than decrease, the importance of differences between places by confronting actors with a more diverse range of locational options. For example, we shall attempt to show how the spatial contexts in which firms have evolved have significantly affected their character, performance and impact, and indeed how spatial considerations have become increasingly critical in firms' competitive strategies. Moreover, internationalization increases the extent to which the fortunes of different places are interdependent.

Another theme concerns the relationships between social, economic, technical and political aspects of industry. Industrial sociology has traditionally suffered from a tendency to underspecify several important economic and technological parameters of the social relations and organization of the firm: for example the particular forms of competition to which the firm is exposed, the technology and labour processes used, the external relations of the firm in terms of the general economic climate, the balance of class forces, and more generally the disciplines which a capitalist economy imposes upon firms. Likewise, research on government–industry relations has often suffered from poor knowledge of industry, technology and product markets. Conversely, work on technological and economic change frequently underestimates the role of the social organization of production in general, and the role of management in particular.

A major attraction of Marxist theory is its insistence on the social relations and material form of production, particularly in the concept of the labour process. Furthermore, recent work on 'spatial divisions of labour', particularly that by Doreen Massey, has advanced understanding of the 'variety and interdependence' which characterizes the space economy (Massey, 1984; Massey and Allen, 1985). Yet the Marxist and Marxist-influenced literature also has several problems; it has little to say on product innovation as a form of competition and, given its primary concern for capital–labour relations, it is weak on how markets function and on intra-industry relations such as subcontracting and various types of collaboration between supplier and buyer firms. Another problem in the Marxist literature is the disproportionate attention given to particular kinds of industry and activity, most obviously routine mass production, especially assembly, and to a particular kind of worker, the blue collar 'mass collective worker'. However, an increasing proportion of the workforce does not fit this stereotype and only a minority of manufacturing employees are involved in mass production; small batch, project and service work are neglected. The net result of these problems is that Marxist and other radical depictions of industry and the international division of labour have become distorted by misleading stereotypes.[3]

More generally, we intend to counter the reductionism which has plagued explanations of industrial development. By this we mean the tendency either to reduce the many concrete forms taken by industry to a small number of aspects identified by abstract theory or to treat particular contingent outcomes (such as runaway industries or deskilling) as the only possible results of capitalist development. To avoid these errors we do not intend to abandon abstract theory but to show how the processes it illuminates are mediated by others to produce a wide range of effects. While we believe that the diverse forms of industrial development are explicable, we do not expect to be able to forecast changes with any reliability, particularly in an industry as turbulent as electronics. As Popper and other philosophers of science appreciated, since social change depends *inter alia* upon the development of human knowledge it cannot be predicted with any certainty because to do so would require us to predict the future content of that knowledge. For this and other reasons, it is now increasingly accepted that social science can only be expected to provide non-predictive explanations, although these can still be helpful for informing practice. In contrast, purely predictive social science, far from being useful for practice, risks hampering it by obscuring the fact that the future depends on what people decide to do: it is therefore more helpful to understand the logic or rationale of existing practice.[4] Since we criticize many popular theories of industrial development precisely for extrapolating transient characteristics into the future, we do not intend to make the same mistake. In researching this book we have repeatedly found the predictions of the most knowledgeable industrial analysts and academic theorists to be highly fallible: one year a certain firm or corporate strategy is hailed as a success, the next it is in serious trouble. This is not a comment on the competence of such analysts, just an observation on the intrinsically unpredictable character of human practice; 'backing a single horse' hardly does justice to its ingenuity and variability. But while we should be surprised if some of the particular patterns we identify below as current are not suspended or even reversed in a few years, what matters is the adequacy of our explanations of what has happened and is happening, and on this we are prepared to stick our necks out.

Organization

Central to our view of the industry is the belief that uneven development involves the articulation of processes which operate at – or rather are manifested at – different spatial scales. For instance, inter-firm competition is most apparent at national and international scales but the recomposition of the workforce is most apparent at the local level. While, for the purposes of presentation, we have found it convenient to deal with international, national and regional or local levels successively, these different aspects are highly interdependent, as we hope to show. There is not simply a single direction of determination moving down from international context to individual firm or plant, for the whole is of course actively constituted by the latter in a way which enables local features to make an impact at larger scales.

The structure of the book is as follows. In part I (chapter 2) we introduce some of the main theoretical perspectives informing and arising from the research. These perspectives are then applied in part II to an analysis of the nature and uneven development of the electronics industry at the international level. By examining the changing nature of four key sectors – semiconductors, consumer electronics, computer systems and telecommunications – we show how the historical development of the international division of labour in the industry has arisen through the mediation of capital accumulation by spatially and historically specific technological, economic, political and institutional characteristics. We conclude part II by refining some of our theoretical points in the light of this empirical analysis.

In part III we examine the British electronics industry as a product of the interaction between this context and certain characteristics of the British social formation. This includes an assessment of the regional organization of the industry within Britain. Arguing that regional development must be seen not only in terms of the characteristics of regions but in the context of wider structures within which they are situated, such as spatial divisions of labour, the main object of part III is to examine this interaction for two radically different regions in which the industry is present. South Wales, a traditional industrial region inseparably associated with 'smokestack industry' and 'miners and militancy', appears to be the antithesis of a 'high-tech' environment and yet has attracted a number of electronics firms. The English M4 corridor, by contrast, appears to be far more congenial and is widely seen as a success story in terms of electronics and has a very different kind of industry. While we assess the impact of the industry on regional development in these two cases we diverge from traditional practice in regional studies by focusing on the nature and performance of the plants in these contrasting areas and on their management–labour relations, so as to assess their significance for regional and national development. Part III concludes with an assessment of the prospects for the industry and innovation in general in Britain under the Thatcherist neoliberal regime. Finally, in part IV we summarize the main theoretical conclusions of the book, and discuss some of the implications for an alternative industrial strategy.

2

Industry and Space
Theoretical Perspectives

No empirical research is ever theoretically neutral, though it may be unaware of its theoretical content. In the belief that it is best to make that content explicit we shall outline some of the theoretical ideas used in our research. However, the theoretical views with which we began the research did not of course simply determine the empirical results but were significantly modified in the light of them, and what follows reflects this process. While the theories we discuss at this point are primarily academic in origin, empirical research also has to engage with the oft-forgotten 'lay-theories' of actors themselves, i.e. the interpretations given and consciously or tacitly used by workers, managers, union officials, business journalists and analysts, politicians and so on. The accounts they offer have a double interest to us: first as part of the object we are studying and second as rival explanations to those offered by academic theories.

We do not propose to review the literature or undertake a systematic critique of rival academic theoretical ideas. Our views on these will be left largely implicit, though occasionally we shall comment briefly on some alternative positions, particularly where they have been highly influential. We shall begin with some methodological points regarding the relationship of theoretical and empirical research, and then present the main theoretical themes.

Methodology

Any attempt to use theoretical ideas consciously and critically in empirical research must appreciate the difference between abstract and concrete research. While such issues may seem rather arcane from the point of view of someone interested in the electronics industry we. have repeatedly found that, without a grasp of this distinction, such efforts are bound to produce 'naive falsifications' and conceptual confusion.[1]

Abstract theories, such as Marx's theory of capital or Alfred Weber's theory of industrial location (Marx, 1976; Weber, 1929), deal with one-sided aspects of their objects of analysis, ignoring others. As such they abstract from numerous other properties of concrete, i.e. many-sided, objects such as IBM in the United States in 1986 or industrial relations in a Japanese television firm.[2] At first sight this distinction may seem unexceptional but what is unusual and often unnoticed about it – with disastrous results – is that it is quite different from and incompatible with another more popular sense in which abstract and concrete are treated as synonyms for ideal or mental and real or factual respectively. While this is not the place to pursue the matter further, basically the problem with such an interpretation is that it sets up a contrast between apparently dispensable theoretical terms and apparently certain and objective factual descriptions – a contrast which the last 20 years of debate in the philosophy of science have shown to be quite untenable. Abstract one-sided concepts are no less – and no more – capable of referring to real objects than concrete ones. The difference between abstract and concrete concepts is not therefore one of the former being less and the latter being more 'connected' to reality, but concerns how many aspects of a particular object are under consideration. Both have fallible connections: both can claim to refer, in some fashion, however imperfectly, to real objects.

What then can abstract theory tell us about concrete subjects like IBM in 1986? A single abstract theory, such as Marx's theory of capital, might tell us a great deal about IBM, as a capitalist firm, but it cannot hope to interpret single-handedly the many-sided specificities of IBM's organization and behaviour – how it has used technology, related to governments, suppliers, competitors, workers, customers and so on. To approach a concrete understanding of the latter we would have to synthesize the interpretations of many abstract analyses, each referring to different aspects of the whole. In other words, the different elements of a concrete situation *mediate* one another. In doing empirical research on actual instances of industrial change we have had to move from the abstract to the concrete and examine the interaction of the various abstract elements. For example, to understand IBM's recent development one might try to relate Marxist notions of time economies to theories of technological change, management and government–industry relations.

Often, apparently divergent theories turn out to be dealing with different aspects or levels of the objects of interest and to be compatible after all; for example, it is not inconsistent to acknowledge both that workers can be defined abstractly as members of a class and that they can be motivated by non-class preoccupations of status and gender.[3] The difficulty comes in checking that theories dealing with different 'levels' or aspects of our objects are indeed compatible and that we attribute particular behaviours to the right predicates of the individuals or institutions concerned, e.g. whether particular forms of industrial resistance are concerned with class or status, or some other phenomenon. In view of the importance of mediations, much of this chapter will be concerned with introducing some of the most important, e.g. technology and the social and institutional forms of capital.

Having dealt with these methodological points we can now move on to outline some key substantive theoretical ideas which inform the research. The most pervasive themes concern the nature of capitalist industry. Although the perspective is at root Marxist, our thinking differs significantly from conventional versions in its emphases, and in ways which have a bearing on the ensuing empirical interpretations. We should also add that the following accounts are not intended as exhaustive accounts of theory for the consumption of purists, but are only meant to be sufficient to aid in the interpretation of the empirical chapters.

Electronics firms as capital

Our first point is deceptively simple: however novel activities such as the production of software and satellites may seem, the electronics industry is still subject to the same pressures and constraints as any capitalist industry, producing not only innovation and growth but frequent market failures and slumps. To remind readers of this we will frequently refer to firms as capitals.

The 'bottom line' in the industry, then, is profitability and though profit goals may vary in magnitude and time horizons – with material effects, as we shall see – everything that happens in the industry is influenced by this fact. Technology is a means to this end, not an end in itself. Nor are people involved in the industry simply as workers, managers, designers etc., i.e. merely as parts of a technical division of labour whose form is dictated only by the technical characteristics of production. Rather they simultaneously enter into class relationships pivoting on the ownership and control of the means of production. While capital can cut employment more easily in some countries than in others, the 'bottom line' is again that capital can hire and fire labour, but not *vice versa*. And of course it is capital, and not labour, which makes investment and location decisions. Similarly, whatever beliefs some may entertain about the 'humanization of work', if and when it happens it is subordinate to the overriding condition of profitability. So like any capitalist industry, electronics production involves a specific mode of social organization, and the social relations have a crucial impact on what happens, firstly in terms of the social relations of production 'writ large' – the division between owners and non-owners of means of production – and second in terms of detailed variants of such relations, e.g. concerning work organiza- tion, which Burawoy terms 'relations *in* production' (Burawoy, 1979).

Given the imagery of 'high tech', it perhaps seems strange to call people such as systems analysts and research scientists 'workers'. Of course we do not want to pretend that such people are no different from blue collar workers, but they do have something very important in common, namely their lack of ownership of means of production, which means that they can only work at the discretion of those who do own them. Certainly some may experience stable employment but this stems from having the good fortune to be in the right job at the right time, not from any rights over the disposal of capital. Even where lifetime employment guarantees exist, these are at the discretion of the employer, and of course dependent on the continued economic survival of the firm. Consequently, whether as the 'bottom line' or as a frequent occurrence, insecurity and vulnerability to unemployment is a reality for some and a real possibility for all. It is also true that professional workers usually have more leverage than lower status workers, both in the job market and within the firm on account of their scarce skills. Whereas less scarce workers can only exercise power collectively, those with specialist knowledge and skills of strategic value to the company, e.g. a research scientist, can exercise it individually. The position of managers who are employees is more ambiguous but the important consideration is that they are still obliged to carry out the *functions* of capital in con- trolling accumulation in its financial, technical and social aspects.[4]

Important though class relations are, the ownership and control of means of production is not only significant differentiation between the personnel of capitalist production. Nor is it the only source of labour's insecurity; indeed we shall argue that other sources have been systematically underestimated. Our argument here rests upon the Marxist distinction between the technical or manufacturing division of labour and the social division of labour. The former concerns the division between individual tasks in the production of a particular commodity under the control of a single capital, say the division of the work of a television plant into different tasks. Here the number of workers allocated to each is determined *a priori* by the firm according to the size of its output and the technical imperatives of conducting the various operations; for example, the operations of assembling, testing and packing may need workers in the ratio of 3:2:1. Within this allocation, individual workers relate to one another as specialized producers of different parts or stages of a particular process of production; they do not each produce commodities individually. Managers and workers of various kinds therefore relate to one another not only in class terms, according to their control over production and over the disposal of profit, but as occupants of different parts of the technical division of labour. By contrast the social division of labour concerns the division between workers in different and independent lines of production, such as potatoes and televisions. Since this is generally under the control of separate capitals it is governed predominantly *a posteriori* by the 'anarchy' of the market, as opposed to the 'despotism' of a particular capitalist; and workers in different firms relate to one another as producers of different commodities.[5]

There are many qualifications to be made to this distinction. The first concerns the relationship between class and the technical division of labour. In principle these two structures could be separated and the hierarchical element of the technical division of labour muted, but under capitalism the technical division itself takes on a class character. Certain kinds of tasks are of strategic significance in controlling production and therefore capital has to be careful not to allow the form of its technical divisions of labour to yield control over the goals and direction of production to its workers. However, just because one set of workers in the technical division has higher status and more skilled work than another set, this does not mean that the first group dominates or controls the destiny of the second. In some respects, therefore, a technical division need not imply class relations and so the two concepts should not be collapsed together from one direction or the other. The second qualification concerns the fact that, while the technical considerations are dominant in shaping the technical division of labour, social influences are usually present too. It is not just that decisions about the kinds of persons to be allocated to each task are influenced by social distinctions, particularly those of gender, but that the design of jobs and division of work itself may be influenced by such distinctions, as in the separation of many administrative tasks, perceived as 'men's work', from related clerical tasks, perceived as 'women's work'. Nevertheless, in describing such matters as 'social' we are using the term in a different sense from that in 'the

social division of labour'. Thirdly, as regards the social division, qualifications are needed (a) to acknowledge that it holds between firms
making competing products as well as different products, and (b) to
cover the situation of large firms making a wide range of different
products where the volume produced of each depends on the market
so that they internalize parts of the social division of labour.[6]

Thus, in explaining the economic insecurity of workers we stress
that it derives not only from the nature of the capital–labour relation
on its own (the fact that workers are non-owners of means of production) but from the nature of capital–capital relations and markets
as well. Social production is largely unplanned at the social level and
is hence the outcome of all the independent decisions of competing
individual and separately owned and controlled capitals. Each firm
cannot afford to wait and see what other firms do but must take
risks in committing capital in the hope of finding profitable outlets
in the market. Furthermore, in thinking about this uncertainty, it is
essential to keep in mind the enormous complexity and distanciation
of markets, particularly in an industry as international as electronics.
It is against this uncertain context that each capital tries to defend and
improve its position. To be sure, the hidden hand of the market, via
price signals, provides a certain degree of coordination and resolution
of conflicts. But it is an *a posteriori* resolution, working behind the
backs of producers as it were. In practice, even overlooking the
divergence of interests between production for profit and production
for use, the mechanism only provides a very imperfect regulation of
the social division of labour. In most sectors, lack of information
on conditions, the unpredictability of what each capital will try to
achieve, plus the lumpiness, rigidity and inertia of so much capital
investment, make it impossible for firms to fine tune and smoothly
adjust their contribution to social production. Instead, serious over-
shooting and undershooting are the norm. The regulation of the
social division of labour is therefore not harmonious and smooth
but discordant and jerky: the hidden hand is decidedly unsteady. For
these reasons, and particularly in the case of electronics, it is therefore
not unreasonable to talk of an element of anarchy in capitalist
competition. So it is not just that workers are vulnerable through
being propertyless, but that the firms that they work for have to con-
duct their business in such a turbulent and uncertain environment.

Competition and the law of value

Any capitalist firm is potentially or actually subject to three sources of
pressure: in product markets from competitors; in investment markets

from firms offering investors higher returns; and from its own workers trying not only to improve pay and conditions but to use the working day as they see fit. In so far as a firm succeeds in responding to these pressures, it renews the same pressures on rival firms.

The types of response are several. The simplest are those which reduce the costs of inputs by finding cheaper sources of materials and cheaper labour. This strategy has been more important in electronics than in many industries because of its combination of elements of low-skill labour-intensive work and high value per unit weight of products which therefore give scope for locating such work in the Third World. A second way, often combined with this, is to lengthen the working day.

More complex, but of greater significance for the long-term development of capital, is the pursuit of endogenous changes in each firm through speed-up, work reorganization, technical change or shifting into new markets. All these, except possibly the last, involve time economies in the circulation of capital.[7] Workers are normally hired for finite periods of time, creditors must be paid off over specific periods and the stream of purchasers and sales over time must relate so as to avoid cash flow problems and provide satisfactory rates of profit. Workers of all kinds are therefore under pressure to reduce idle time during the working day and maximize productivity. Time economies are relevant to each stage of the circuit of industrial capital, not just to direct production but to R&D, marketing, customer service and administration. In fact in some cases there may be greater scope for their realization in, say, the processing of invoices and the authorization of expenditures than in the processing of materials. Similarly, as we shall see in the computer business, time economies in customer service not only foster goodwill but allow for more rapid turnover of capital and hence greater profitability.

The concept of time economies refers to the effects of *process innovations*, be they technical or organizational. Yet despite their undoubted importance, on their own they constitute only half the story of how firms compete and survive. The oft-forgotten other side – in both neoclassical and Marxist accounts – is that firms also compete through product innovation and seeking out new markets, which can provide new opportunities for the production and realization of surplus value.[8] This is the main type of competition across large areas of the electronics industry. Nor is price always the main consideration in competition; in defence electronics, for example, a premium is often placed on quality and performance rather than price. Competitive strategies based solely on cost-lowering through process innovation may fail dramatically in the face of successful product innovation: there is little economic gain for capital in being able to produce an obsolete commodity with unmatched efficiency.[9] It is also in the interest of firms to try to transfer their operations into the production of commodities with a higher value-added than existing ones, and product innovations which enable this will be preferred to those which do not.

Loosely, 'the law of value' refers to the process by which a continually increasing pressure bears down upon firms, forcing them to reduce the labour time involved in commodity production to state-of-the-art levels and to seek new sources of surplus value; in turn it obliges other firms to do the same, on pain of extinction. Ultimately it is this, and not some inherent dynamism in the technology of electronics, which drives the industry. Certainly the technology of electronics offers exceptionally abundant *possibilities* for development of process and product innovations, but this is a permissive factor rather than a driving force, in the sense that it offers less resistance to the pressure of the law of value than more technically restricted industries. This may seem a fine point but confusion on this issue supports wider misconceptions about the expansionary potential of high-technology industry.

If competition is the whip hand behind the development of industry it also has its costs. Unless there is very rapid market growth or some form of limitation on entry into the market, competition invariably involves not only winners but losers, whether as a positive-sum game in which there are some absolute losses and losers, a zero-sum game, or, in the case of declining markets, a negative-sum game. Putting this last point together with that concerning the unevenness of market regulation, it becomes clear that market failure, as much as success, is endemic. Therefore, a concrete event such as the collapse of a microcomputer firm must be seen not as an aberration, or simply as a failure of an individual capitalist, but as a structural inevitability in any capitalist industry. (However, which *particular* firm fails or succeeds can be explained in more individual and specific terms.)

Uneven Development

'Uneven development' is now a widely used term in the regional development literature, extending beyond its original Marxist roots.[10] Perhaps inevitably, its meaning has lost some of its specificity, so we must clarify our usage of it. By uneven development we mean not only uneven spatial development, but the uneven *processes* of development endemic in the structure of capitalism itself. These processes differ from those which might exist under a different mode of production. First, there is an unevenness or asymmetry in economic power which derives from the concentration of control and ownership of the means of production into the hands of a minority and the consequent class nature of society. Second, the relationship between the market signals – especially profit rates – which guide capital and the social needs for goods and services is also highly uneven, such that needs are met only in the most approximate fashion and in terms which suit capital. For this reason, products such as low-income housing and public transport are neglected. Third, the unplanned nature of development and the anarchy of competition make the process of capital accumulation uneven. Fourth, the rates at which time economies and other forms of cost savings are made is inevitably uneven, given the differing technological possibilities open to different activities. While the latter type of unevenness might be expected in any kind of economy, in capitalism the response to it is not shaped by evaluations (of whatever kind) of needs but by the resulting shifts in the pattern of variation of existing and expected profit rates across the economy. The resulting continual flux of capital from unprofitable activities to highly profitable ones can create sudden unplanned reversals and switches in development, and there is no mechanism to ensure that new investment is forthcoming in the right amounts and at the right times and places to ensure continuity or orderliness, let alone according to criteria of need. In short, disproportionalities, booms and slumps, boom areas and slump areas, are part of the normal functioning of capital and certainly not merely the result of exogenous disturbances.

Yet while the development of activities and areas is uneven, they are not separate but interdependent or 'combined'. For example, it is vital to appreciate the way in which the development of the likes of Silicon Valley and the Boston 128 corridor is actually linked to the very different kinds of activity found in places as diverse as central Scotland or Malaysia. Indeed, apart from the failure to look at electronics as a capitalist industry, the widespread ignorance of the combined nature of the development of its parts and localities is perhaps the most serious fault in both popular and academic accounts.

Technology

A further and crucial mediating element of industrial development is technology. While particular technologies, such as those of electronics, can be used in many different ways, they have specific capabilities which limit this variety and restrict development along 'natural trajectories' (Nelson and Winter, 1977). The possibilities are limited not only by human ignorance but by the properties of the objects themselves. Yet both neoclassical economics, which reduces technique to the outcome of price ratios, and certain variants of Marxist theory of the labour process which treat technique as a function of class struggle create a false picture of technology as plastic and hence as an insignificant aspect of development (Walker, 1985b). There is no contradiction in acknowledging both that the technical properties of materials (including the powers of people themselves) have an irreducible quality, and that their use and manipulation is always influenced by social characteristics, though in ways which are in accordance with the natural laws obtaining for the materials. For instance, the design of computer systems and the technology used to produce them are influenced by features of the social organization of their production, particularly in terms of the relative power and remuneration of different occupations and the pressure to produce the systems profitably.

These constraints and pressures deriving from technical possibilities often extend beyond the production of individual commodities to whole families of products, among which there are *technological complementarities* or *synergies*. These occur where one kind of technological development enables or presupposes others or where strength (or weakness) in one field of production makes for strength (or weakness) in another. On a micro scale, one kind of production may be physically impossible without another; so, for example, microcomputers require microchips, which in turn require silicon substrates, and the design of the latest microchips is itself dependent on the existence of chip-based computing systems. On a macro scale there are synergies between groups of products and production processes; for example, strength in microelectronic components facilitates the development of electronic capital goods and computers. This implies that the whole of the development process is more than the sum of the parts. Because the indigenuous consumer electronics industry in Britain is so small and relatively backward, electronics components firms lack a mass market and hence are denied opportunities open to Japanese producers. Conversely, the consumer industry in Britain is inhibited by the weakness of the indigenous components industry.

Sometimes there are barriers rather than synergies. Where different kinds of production require radically different approaches, firms which are strong in one may find it extremely difficult to transfer to the other. The gulf between defence electronics, especially where it involves project or very small batch work, and consumer electronics and commercial computer systems is an outstanding example of this. In the former case, price and process efficiency are relatively unimportant and products have to be both highly specified and ruggedized; but success in the civilian markets requires not only different products but quite different management skills, particularly in low-cost volume production and marketing.

Research and Development

R&D has gained a special significance as a particularly prized economic activity from the point of view of regional and national economic development. However, such views sometimes fail to appreciate that its economic effects in terms of feeding innovation and economic development are neither direct nor necessarily geographically localized and congruent (Sayer, 1983). Further, there is a tendency to treat R&D and innovation themselves as simple categories. Innovations are too often assumed to consist of discrete breakthroughs (the 'big bang' theory of innovation) rather than as part of a continuous stream of development. Most R&D is development work and what fundamental research is done always builds upon previous work. This is not to say that there is little difference between the two: on the contrary, fundamental research is a far cry from much of development work, which often consists of modifying designs for particular markets. Nor is it to belittle development work in general, for the process of turning an invention into a product which can be sold competitively and at a profit is the more time-consuming, expensive and critical side of R&D. Moreover, plenty of process and product development goes on outside R&D laboratories, in production, often through learning-by-doing (see Rosenberg, 1982; Walker, 1985b). The work of engineers involves a design element and software 'production' is substantially software design and development. Finally, a significant proportion of innovation goes on in user firms, often in collaboration with the supplier, as in the case of large banks modifying computer systems (von Hippel, 1977; Storper and Walker, forthcoming). Unfortunately, it is precisely such matters which are overlooked in analyses which simply count up discrete innovations and R&D centres and read off economic conclusions from the results.

Markets and marketing

There are also some common misconceptions about the nature of markets and demand which need countering if the activities of industry are to be understood.

For the purpose of abstract economic theory, it may be justifiable to reduce marketing and exchange to the transfer of the title of ownership of a commodity from producer to customer, and even to abstract from the problems of imperfect information, from the material organization and practices of markets and from the difficulties of distribution and communication created by the distanciation of producers and consumers which we mentioned earlier. But in a concrete study one simply cannot ignore the kinds of work involved in capitalist industry in coping with these problems – after all, there is a labour process involved here, just as much as on an assembly line. Contrary to the impression frequently given by both neoclassical and Marxist economics, products do not sell themselves to users the nature of whose demands are already given, under conditions of perfect information. Markets are material organizations and sets of practices (see Storper and Walker, forthcoming). In an industry like electronics most markets are other firms or public institutions. The fact that the price mechanism may operate behind actors' backs, as it were, through the 'hidden hand', does not mean that exchange takes place out of thin air, without work or organization. Most wants or needs are not innate properties of individuals, already specified and merely awaiting a supply to be satisfied. Needs are strongly influenced by existing patterns of activity or 'productive consumption', their configurations, technical specifications and associated organizational forms and work routines. Although for needs to be revealed in the market there must also be purchasing power, this should not allow us to forget these use-value constraints. Technical specifications in particular are enormously important in marketing and competition in information technology. So, for example, selling a particular software package is not merely a question of finding users with money, but finding ones whose existing set-ups (operating systems, hardware, operator skills) are suitable. Moreover, it is important to note that wants and needs evolve through an interaction between producer and consumer – though as we shall see the type and intensity of interaction vary considerably. In all cases, pure exchange – the transfer of titles of ownership – is set within processes of information exchange and negotiation. In some cases, there are associated collaborative relationships, while in others where political influences are strong, e.g. military equipment and public telecommunications markets, there is a critical process of political lobbying between buyer and seller.

Another reason for a dismissive attitude towards marketing lies in the tendency to assume that all commodities are like familiar mass consumer products. In radical thinking it is often reduced to advertising on the model of that used for products which are thoroughly familiar, i.e. a redundant zero- or negative-sum game among producers involving a net wastage of resources to society. Unfortunately, the justifiable disdain felt on the Left for this kind of marketing, together with arguments that the use values are sometimes illusory, and reinforced by antipathy towards the types of people who do it, has undoubtedly prompted many to ignore marketing altogether as a significant phenomenon.

Certainly in the consumer case, commodities such as radios and soap powder are either familiar or easy to learn how to use, so marketing is virtually reducible to advertising and involves a largely one-way flow of information from producer to consumer. Yet even here the distanciation of producers and consumers can create

Mass production for consumer market:-

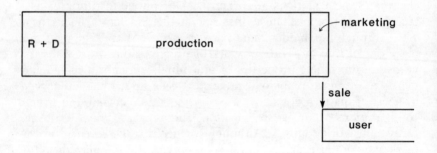

Custom or project production for business market:-

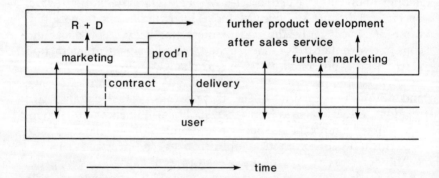

Figure 2.1 Supplier–user relationships in two contrasting markets

problems in the successful realization of the value of commodities in their sale which make marketing and distribution important sites of competition. By contrast, many goods sold to industrial or institutional customers are difficult to use, e.g. air traffic control systems or telephone exchanges. This is particularly so for those which are to any extent customized and in which product innovation is common. Here, marketing involves a two-way process of communication between producer and user, who jointly define what is to be produced and sold. In either case, the market is not simply 'out there' but has to be sought out and created, particularly where new products are concerned. As Walker (1985b) comments, the 'sales effort; is not mere ideological manipulation but a necessary condition for, and a significant part of, the development of industry.' Figure 2.1 illustrates the difference in buyer–seller relationships for standard mass consumer products, such as televisions, and for highly customized products for industrial or public sector use, such as missile guidance systems.[11] It also suggests the implications for relationships between marketing, production and R&D.

A further implication of these points is that, strictly speaking, market demand is not an acceptable explanation of the course of technological development, *and vice versa*, for both omit the crucial interactions between producers and users in the sphere of marketing, coupled with development and after-sales feedback and support, through which products evolve. Therefore arguments about whether certain developments arose through 'technology push' or 'market pull' are at best crude simplifications (Walker, 1985b).

The labour process and management–labour relations

If we are not to treat industry as a 'black box' we need to consider how its labour processes are organized and what kinds of relations exist between management and labour. These matters are central to the character of particular industries, to the competitiveness of individual firms and to the kinds of work, workers and experiences of work in the industry. While we have not looked at individual plants or offices in sufficient detail to analyse the labour process in depth, in part III we shall comment on the general forms of work organization and their associated management–labour relations.

Interest in the labour process was largely awakened by Braverman's eloquent but flawed *Labor and Monopoly Capital* and this has coloured the way industry as a whole is seen by many on the Left (Braverman, 1974). On this view, capital's attempt to control labour is elevated from the level of one among several means to the end of profit (albeit a very important one) to an end in itself. All changes in the labour process are seen as caused by this and particular forms such as 'Taylorism' or 'Fordism' are interpreted as unmediated expressions of the logic of capitalist development.

Subsequently, Braverman's early critics reinstated labour and class struggle as active forces affecting the labour process but continued to treat control as a zero-sum affair and as the only reason for change, ignoring product-market and labour-market influences. We do not wish to deny the place of struggle in the labour process, but labour processes are often changed because the available process technology and/or nature of the product have changed rather than as a way of winning greater control over labour for its own sake. In fact, capital may sometimes replace a labour process over which it has great control by one over which it has *less* control, if the new process is the only existing way of producing a new commodity for which market prospects are better than for its predecessor. Technology is rarely 'plastic' in the hands of management; it can create rather than solve problems of control. Therefore an increase in labour's control over work may be a 'windfall' gain rather than a hard-won victory. And similarly for capital, if its power over labour is increased. In other words the power of either side can be affected by change originating *outside* the workplace in terms of changes of markets for products (Kelly, 1982). Intensification of labour and deskilling are not the only means towards the goal of profit, nor the only possible result: product diversification and cheapening of constant capital are alternatives. So new technology may or may not have the effect of weakening labour's control over production, and when that effect does occur it may no be the main reason for its introduction.[12]

Certainly an explanation of changing work organization which puts all the weight on the historic struggle between capital and labour sounds more radical – until, that is, you notice how this gives the impression that labour must lose in the long run as the forces of production develop and capital inexorably deskills labour. But a radical ring is no substitute for theoretical rigour.

These distortions receive support from the habit of neglecting product innovation which was mentioned earlier for, without new products, net deskilling in the long run seems more credible, though even new process technology need not necessarily involve net deskilling. The issue of how skill levels change is therefore emphatically an empirical question. In electronics, it happens that instances of upgrading of skills are plentiful and skill shortages are common despite high unemployment.

Other recent developments in labour process theory include a more receptive attitude to ideas of bureaucratic control and the significance of consent and motivation, together with a realization that naked coercion is an ineffective way of securing high productivity (Burawoy, 1979; Littler and Salaman, 1984). Research in this area has become more empirical and more alive to the fact that particular forms of organization are not simple reflections of the state of development of industrial capitalism but are heavily conditioned by their situation in particular local conjunctures. So, for example, differences in the organization of the labour process in the West and Japan reflect those local conjunctures, and though some may diffuse and be imitated – as we show in chapters 10 and 12 – they invariably have to be modified to suit local conditions.[13]

Finally, another important development lies in the recognition that while firms are primarily economic units, their internal structures bear the marks of, and indeed actively reproduce, other influences, such as gender and status relations. These influences have an 'expressive' side to them – for example they concern how individuals are regarded by others and how they see themselves – and hence they affect workers' motivation and identification with their companies (see Burawoy, 1979; Cockburn, 1983).

Social and institutional forms of capital

Although the electronics industry is structured at the most basic level by capitalist social relations of production, this allows room for a wide range of particular social and institutional forms of organization within this structure. For example, no concrete study of industry could fail to be impressed by the striking differences between British and Japanese capital in the way they are organized and in the consciousness of their workers. Such variations are interesting in themselves, but their primary significance for us lies in the fact that they make a difference to the competitiveness of firms and help to explain uneven development. So while it may be acceptable to say, for the purpose of abstract theory, that 'capitalists behave like capitalists wherever they are' (Harvey, 1982, p. 424), this will not do for explaining concrete patterns of competition and development.

The variations come out across a range of areas, many of them overlapping – in management–labour relations, 'work culture', gender and racial divisions, authority relations, ways of organizing labour processes and using technology, commitment to technical skills, corporate structure. Most obviously, the variations relate to the kind of activities carried out; for example, it has been argued that in the relatively stable environments provided by mature consumer markets, internal efficiency is the key, while for firms needing to develop new products continually in order to survive in unstable environments, 'effectiveness' and flexibility in meeting external challenges are more important (Burns and Stalker, 1969; Newman and Newman, 1985). All these internal characteristics are strongly related to local circumstances outside the firm, e.g. regarding relations with the government and financial capital, labour markets and employment legislation and the conditioning effect of wider cultural characteristics. Usually these mediating influences on capital are taken for granted or are only noticed where they differ strongly from those of our own society. So while it is common for Western commentators to note the influence of cultural factors on Japanese industry (albeit usually in a gestural fashion which obscures more than it explains), few notice that it is ethnocentric to suppose that Western capital is not equally influenced by its own social and cultural context.

Yet these social and institutional forms of capital do not merely represent a kind of environmental conditioning by the local societies in which capital is situated; they are also partly consciously cultivated. This is most obvious in the American expression 'corporate culture', meaning a set of institutional norms regarding individual and collective behaviour in the firm. However sceptical British readers might feel about such neologisms, it is important to appreciate that they are more than mere 'public relations'. As we shall see in chapters 10 and 12, corporate cultures make a difference to the efficiency and competitiveness of firms, and where they are consciously cultivated it is for that purpose.

As the social forms are moulded by the local conjunctures in which they develop, it may therefore be difficult to change them *in situ* or replicate them in a new location. But multinational firms can take advantage of differences in local societies by locating operations with different requirements in different areas. Since productivity and competitiveness are affected by the social forms of capital, then firms may command an advantage – or suffer a disadvantage – by virtue of their location. Moreover, it means that the establishment of new plants always requires *mutual* adjustment between company norms and established local practices. Workers are not born already fashioned to corporate specifications; they have to be made, both in the workplace and, more problematically, in the community. Sometimes firms may move operations primarily to escape the restrictions of old practices associated with their existing locations and to establish new ones. This is usually the dominant consideration in the preference for greenfield sites. It is multinationals which are usually the key bearers of new organizational forms, carrying them into new areas and thereby changing local work cultures. We shall look at examples of this process of adjustment in chapters 10 and 12.

An important implication of this is that the evaluation criteria for different locations are by no means limited to the costs of inputs into production, such as labour and raw material costs; often a far more important consideration is how effectively the inputs can be combined in the workplace, which is largely a problem of the quality of labour, management and social organization. As we shall see, this is increasingly recognized by governments and agencies trying to attract or support capital in particular places. In explaining the shifting patterns of international competition it is therefore necessary to consider the comparative advantages of the social characteristics of the countries and regions in which the firms operate.

The nation state

One of the most neglected aspects of uneven development, even in studies at the international level, is the role of nation states. Too often, the internationalization of capital is treated as if it took place in a stateless *laissez–faire* world economy; in effect, such studies fail to register that it is literally *inter–national*.

To do justice to the role of nation states, we must recognize their double nature as context and agent. Capital operates in different national social formations, each with its own historically developed social characteristics, providing particular kinds of economic contexts in which capital can sell and produce and specific political, fiscal and legal contexts supporting and regulating capitalist production and the reproduction of the workforce (see Burawoy, 1983).

But states are not just passive, albeit differentiated, containers within which capital and labour produce uneven development; they are also agents within that process. They actively influence capital accumulation and international competition through their support for domestic capital and their stance towards foreign capital, whether the latter is based within or outside the country. Through macroeconomic policies (e.g. regarding balance of payments problems) they respond to the net effects of the myriad of capitals operating within and outside their jurisdiction. Through defensive protectionism, such as imposing tariffs and 'voluntary' restrictions on imports, governments alter the balance of international competitiveness and profoundly influence patterns of the internationalization of capital.

For example, protectionism is the main reason for Japanese investment in consumer electronics in Europe and a major factor in
American investment in semiconductor plants in Europe. However,
this influence is widely underestimated and is rarely treated as being
of any theoretical interest. Often, defensive protectionism shades
into aggressive nationalism, usually involving corporatist relationships with leading domestic producers. This may take the form of
funding and coordinating R&D, simple subsidies, nationalization,
encouraging restructuring or mergers to achieve economies of scale
and scope, imposing technical standards which keep out foreign
competitors, adopting preferential purchasing policies and so on.
The state's role as a major purchaser (particularly for computers,
telecommunications and defence equipment) is also an important
lever in such intervention; so, for example, computer firms investing
abroad are unlikely to risk antagonizing host governments too much.
All of these types of intervention have again had a profound effect
on international competition, perhaps nowhere more than in Japan.

This is not to say that such policies always have their desired
effects. On the contrary, they sometimes not only fail but produce
the opposite of the desired results. The most common case is
perhaps where protectionism, which is intended to allow indigenous
capital to become strong enough to compete internationally, allows
it to become less competitive than before.

There are also other kinds of state policies, which though not
intended to influence the international competitiveness of domestic
capital, nevertheless do. For example, by setting highly specific
technical requirements for defence and telecomnunications products,
the state, as a prime purchaser, may inadvertently restrict overseas
sales of those products by its leading firms.

The effects of the role of states in the process of international
competition are rarely straightforward, thanks to the growing non-
correspondence of capitals and nation states, as internationalization
proceeds. Contrary to some early radical views, this non-correspondence certainly has not simply freed multinational capital from
dependent relationships with states. Even the most international of
firms still have a heavy bias towards their country of origin and,
although the Left likes to emphasize cases where multinational
companies have had the upper hand over host governments, the latter
are by no means powerless. The relationship is never one of simple
one-way dependence or domination. Capital needs state support and
is to some extent therefore in a dependent position, and there are
limits to its ability to 'shop around' among countries, most obviously
where it needs to penetrate particular national markets. So the rise of

multinationals has changed the way in which states relate to capital but has certainly not relegated states to a minor supporting role.

Yet posing the issues in this way still hardly does justice to the complexities of the field of possible competitive, contractual or collaborative relationships which can exist. We can begin to appreciate them by considering the ambiguity of the terms 'domestic capital' and 'national economy', which could be taken to refer to the following:

1 domestically owned capital located in the home country;
2 domestically owned capital, including that located abroad;
3 all capital, of whatever nationality of ownership, operating within the country in question.

Such problems are by no means of purely academic interest, but are vital to policy questions, for both states and firms, e.g. IBM's claims to be a domestic producer in France and Britain. In detail, a highly internationlized and politically important industry like electronics is structured by the following sorts of relationships:

1 between capitals of the same nationality;
2 between capitals of different nationalities;
3 between capitals and their home governments;
4 between capitals and foreign governments;
5 between governments of different countries.

Each one of these can involve both identities and conflicts of interest, and numerous situations can arise where several of these pairs combine and cross-cut. Hence there is much more to the implications of the non-correspondence of capital and nation states than a simple opposition between the two. Moreover, power relationships between big capital and governments do not always involve a simple opposition of independent entities, and multinationals appear to seek government protection or assistance no less than smaller national firms. For example, where the state acts as buyer or as adjudicator on technical standards, it must be remembered that in doing so it is generally guided by the advice and expertise of the most powerful supplier firms. However, this does not rule out the possibility of the state failing to do what is 'functional' for capital.

Given the way the relationships we have listed cross-cut one another, the field of possibilities facing states and firms for various types of strategic action is very large. In any competitive situation, 'beat 'em or join 'em' decisions frequently have to be made, but in an internationlized industry the costs and benefits of competition and collaboration are affected by differences in nationalities of ownership and by the relationships of multinationals to host governments.

Simple collaboration between firms of different nationality may be blocked by governments on the grounds that they are against the national interest. (For example, a major link between IBM and British Telecom was blocked after protests from the indigenous computer industry.) Simple state support for domestic capital as a way of countering foreign competition is complicated by the presence of often large foreign-owned competitors as 'domestic' producers, and by the presence of conflicting collaborative deals between domestic and foreign capital. So, for example, defensive EEC policies often find themselves supporting IBM. Similarly, where a country restricts imports, it runs the danger not only of inducing retaliation but of damaging the interests of 'its own' multinationals abroad, where they are responsible for the imports, and of penalizing indigenous firms which are dependent on the use of imported technology.

The picture is further complicated by the existing pattern of uneven development in which capitals and states are situated and in particular, in the case of electronics, by relationships of technological dominance and dependence. As we shall see, such a complex decision field invites both strategic action and opportunism, as when a British firm makes deals with Japanese rivals while collaborating with other EEC firms to counter the Japanese threat. Yet however complex, it is still a *structured* decision field, defined not only by capital (and labour) but by nationality and governments.

Space and spatial divisions of labour

We have already argued that space makes a difference to industrial development and therefore cannot be considered simply as an incidental output. However, though many now accept this, some problems remain. One concerns the common habit of distinguishing between spatial and non-spatial processes, e.g. between apparently non-spatial investment decisions and obviously spatial location decisions. The problem here is that *all* social processes are spatial; they are made by, involve and affect particular people and things in particular places whose spatial contexts make a difference. Certainly some processes and institutions have relatively clear spatial expression (local labour markets) while others (product markets) are more fragmented and difficult to identify geographically. Nonetheless, the latter are still spatial.

One of the most fertile concepts in the explanation of uneven spatial development in recent years has been that of 'spatial divisions of labour' (Massey, 1984). Although chiefly associated with the subnational scale, it is compatible with the concept of the international division of labour as used by theorists such as Stephen Hymer (1972). Its chief use is firstly in helping us appreciate how the development of particular localities, regions or nations is differentiated yet interdependent, and secondly in theorizing how regional development relates to the organizational structure of industry. Following Massey (1979) we can distinguish between spatial divisions of labour which are sectorally based, with individual industrial sectors being concentrated in particular regions (cotton textiles in Lancashire, shipbuilding on Clydeside and Tyneside, metalworking in the Midlands etc.) and divisions which are hierarchically based, with different activities or stages of production within each sector located in different regions (headquarters in London, routine office work in the outer suburbs, R&D in the outer metropolitan area, routine production in the north etc.). This can in turn be related to the distinction introduced earlier between the technical and social divisions of labour. The hierarchical spatial division corresponds roughly to the technical division, where the regions are characterized by their different roles within corporate technical divisions, while the sectoral spatial division of labour corresponds to a situation in which regions are distinguished primarily by their role in different parts of the social division of labour.

In practice, the term 'spatial division of labour' has been used rather loosely, at worst meaning little more than 'the geography of employment' and more commonly by authors failing to clarify whether it concerns corporate spatial divisions of labour within individual firms or on an aggregate level corresponding to the social division of labour. Clearly, where there are no multiplant firms, the aggregate social division can exist without corporate spatial divisions, as it did in the days before multiplant firms in early capitalism; but as industrial concentration proceeds the former comes to be cross-cut by the latter. However, progressive vertical integration and hence extension of corporate spatial divisions of labour is not the only option for capital (Scott, 1983) and so even within one sector we cannot expect to read off the aggregate type of spatial division from the first, by assuming that the corporate spatial division of labour typical of the largest firms in the sector represents the sector as a whole.

Despite its undoubted utility, there are several dangers in the way that this concept has frequently been used in the explanation of uneven spatial development. Firstly, in so far as the corporate/hierarchical sense of spatial division of labour (as a technical division) has become the best-known usage of the term (see Massey (1984) for instance), we must not allow this to obscure they way in which uneven spatial development is structured by the state of play of competition between *many* firms in different places. As we shall see in our review of the international electronics industry, this is often the dominant consideration from the corporate point of view. Secondly, each corporate division of labour may consist of several technical divisions of labour. Thirdly, while there is invariably an element of hierarchical control within corporate divisions of labour, there is also frequently a matrix-like element to the corporate structure. Fourthly, the significance of the concept of spatial divisions of labour has sometimes been diluted by restricting it to a 'top down' view in which places appear to be the passive recipients of roles allocated to them by firms. But while multiplant firms can benefit from the advantages of doing different things in different places, it is easy to overlook the way in which the work carried out in different places actively constitutes the spatial division of labour. Moreover, as Massey (1984) emphasizes, geographical variations and forms actually influence the general character of industries themselves. For example, were it not for the proximity of cheap Mexican labour to the American market, many firms would not have split off their labour-intensive activities from their capital-intensive operations. Note, however, that there are hidden costs involved in spatial separation, as some Western firms are discovering (Office of Technology Assessment, 1986); the decentralization of production has helped to divorce R&D from manufacturing and this has led to a decline in manufacturing control and expertise, relative to Japanese firms which tend to favour greater centralization and inter-activity coordination.

A final problem with many uses of the spatial division of labour concept involves a tendency to overlook its abstract character, for it can only function as a *partial* explanation of uneven spatial development. The concept focuses primarily on capital's need for workforces of particular types in the context of different kinds of production activity. Unfortunately, the radical literature has become transfixed with labour as the only significant locational factor. But there are other considerations in location decisions, in particular, getting access to markets. Especially where the product is a bespoke or custom one, there is often a need to locate close to customers, e.g. software houses locating near major public sector users or specialized arms manufacturers near military establishments. Conversely, access to suppliers and subcontractors may be the dominant consideration in location. In radical work, failure to note such considerations is again symptomatic of the neglect of markets, or a tacit assumption that all markets are like dispersed consumer mass markets.

Mediations and stereotypes

So far we have introduced some important types of mediating and contextual influences upon capital accumulation which make significant differences to its concrete forms. In practice all these elements mediate each other. While, for the purposes of analysis, we may start with the most basic or general abstractions and then move on to the mediations, it is a mistake to imagine that such a procedure reflects the chronology of a real process. Capital is not born naked later to be clothed in various mediations; rather its very conception is affected by them and usually its subsequent development carries their birthmarks.

As we suggested at the beginning of the chapter, many of the most pervasive problems in the analysis of industry and uneven development stem from misunderstandings of the relationship between abstract and concrete. The overextension of the concept of spatial divisions of labour was just one such example of this. Researchers have repeatedly underestimated the extent to which the characteristics and mechanisms of capital are mediated by different technologies, forms of social organization, geographical variations and by relationships with nation states and their governments. Consequently, the variety of possible concrete forms of development has been widely underestimated or else seen as the playing out of an unchanging repertoire of responses. Alternatively, particular concrete patterns have often been read as the unmediated expression of the universal developmental logic of capital accumulation in the abstract and therefore endowed with epochal significance; *The New International Division of Labour* (Fröbel et al., 1979) is a good example of this with regard to the runaway industry phenomenon – the flight of certain kinds of manufacturing to Third World countries. Even if such trends or patterns are seen as outcomes of particular global conjunctures, their homogeneity and permanence is invariably overestimated. It then becomes difficult to appreciate how such developments could change, or even be reversed, as the mediating influences alter.

The product cycle model suffers from a similar problem: in freezing a particular spatio-temporal sequence and treating it as universal, it is subject to 'death by a thousand qualifications'; for example, many products never reach a mass market and never become standardized, and foreign direct investment is not always necessary for conquering foreign markets, as Japanese companies have shown. It is also ironically ahistorical for like all stage models it assumes that the same patterns can be reproduced regardless of changing context.[14]

Models based on historical sequences frequently rely on an anthropomorphic metaphor of ageing ('infant industry', 'mature product' etc.). While this is often quite harmless it can give the false impression that certain forms of organization are restricted to particular periods or have a built-in obsolescence. For example, Marx, and later Braverman, assumed that subcontracting was tied to early capitalism and would soon be eliminated by an inexorable process of vertical integration. Likewise homeworking has similar historical associations. However, the conditions in which such practices can thrive are highly varied and there is little reason to suppose that some of them could not persist or re-emerge as capitalism developed. Conversely, there is no reason why progressive vertical integration should continue without reversal. The same arguments apply to the ahistorical view that a focus on what appear to be the leading most advanced companies (itself a questionable focus) will reveal what is shortly to be generalized to the rest of industry.

In other words, the field is plagued by stereotypes and stylized facts – of 'runaway industries', 'branch plant economies', 'external control', cheap deskilled female labour and so on. These always have some basis in real events but they have induced a blindness to instances of different processes, e.g. to the persistence of agglomeration tendencies (now being rediscovered), the immobility and indispensability of many branch plants, and the rise of new skills.

In attaching such significance to particular empirical outcomes, these radical theorists depart from the practice of both neoclassical and Marx's economics, for one of the few things that these conflicting bodies of theory have in common is a view of economic decisions in terms of trade-offs rather than single outcomes. Economic theory may set out the variables affecting a decision, but within certain limits the values of those variables can only be determined empirically, and hence attempts to predict a single species of outcome are bound to be falsified in practice. This is most obviously the case regarding the economics of spatial concentration and dispersion in the location of industry; so much depends on contingent factors such as existing spatial distributions and the technical characteristics of particular industries that prophets of universal agglomeration or dispersion are bound to be partly wrong. Likewise, it would be foolish to prophesy that 'the capitalist labour process' (as if there were only one!) is moving in a single direction, perhaps towards giving workers 'relative autonomy' or whatever, when such changes always depend on a rich variety of contingent conditions.

Specific biases run through these stereotyped views: an emphasis of manufacturing over services; production, especially mass production and assembly, over other kinds of production or non-production activities; consumer markets over industrial markets; labour over management; blue collar over white collar; multinationals over smaller firms; dispersion over agglomeration and so on. These biases are not random; in part they have their roots in the political priorities of radical researchers. But whatever their source and whatever our own political preoccupations, we need not be reduced to choosing between rival accounts of industry according to which one shows capitalism in the worst (or best) light.

'High tech' and related ideologies

One source of misleading ideas having its origins outside academic circles is what we call the ideology of 'high tech'. High tech has become an immensely popular term and some researchers have been tempted to make it into an analytical category, though with little success.[15] However, while this concept is far too incoherent to be of any analytical use its ideological significance is too great to ignore as an object of explanation.

High tech is the embodiment of modernity, high science and technological wizardry. Such terms have had many precedents in the history of capitalism (e.g. 'the age of steam'), all of them suggestive of a 'technical fix' for economic problems, an industry apparently capable of transcending normal economic forces. Of course high tech is tied to specifically modern signifiers, in particular the chip and the computer – enormously complex, yet silent, clean and effortless, and perhaps intelligent or 'smart'. And the technology of the principal signifiers is indeed extraordinary and wondrous, but the key to the ideology of high tech is the way in which the powers and values ascribed to the technology spill over onto anything associated with it. This is surely why commodities as diverse as running shoes, chocolate bars and furniture have been blessed with the name high tech. More seriously, the meritorious qualities of the technology come to rest on its economic and social coordinates, so that high tech industry such as electronics must be an unqualified 'good' for everyone. Indeed, wonderment at the technology seems to have the effect of suspending peoples' critical faculties. Not surprisingly, arms firms, which make up a large minority of the electronics industry, have used the term to great effect, for, while people might have doubts about having a 'missile factory' in their locality, a 'high-tech firm' sounds far more appealing.[16]

A further important construct in this ideological discourse is the 'sunrise/sunset industry' contrast. High tech is of course the epitome of sunrise industry, apparently recession-proof and subject to irresistible growth. Conversely, the prefix sunset is sufficient to condemn a vast range of industries to equally irresistible decline, and hence neglect, regardless of the extent to which their products are still needed (and not merely realizable in terms of contingent effective demand), and regardless of the possibilities for rejuvenation. All the 'rustbelts' can do is try to get some sunrise industry. In other words, such terms remove a whole range of decisions regarding the shape of economic development from the political sphere and fetishize the contingent products of human action as inevitable external forces to which we have no option but to submit. The contrast is further compounded by the distinction between the 'brownfield' sites of the sunset industry and the 'greenfield' sites of the sunrise, the implication being that the environmental effects of the latter could only be benign. In other words, innocuous though such terms may seem, they encourage blind submission to the caprice of capitalist uneven development. As Barthes put it, ideology is 'depoliticised speech'.

Like any ideology, the rhetoric of high tech is forged out of the pre-existing ideological materials. In this case, there is a potent union of popular mythologies of science ('whizz-kids' and white-coated scientists making 'breakthroughs') and of free enterprise (heroic thrusting entrepreneurs, unfettered by corporate or bureaucratic ties). Yet ideologies can also generally find a modicum of apparent empirical support. For example, while those who have gained riches in microelectronics did not generally start from rags, some of the best-known firms have grown at outstanding rates from small beginnings, and have been founded by young scientist/entrepreneurs, the most famous cases being Hewlett-Packard and Apple, both of which originated in their founders' garages in that mecca of high tech, Silicon Valley, California. On the other hand, old firms like IBM have more power than these and in Japan large firms have been highly innovative. And contrary to claims of free market ideologists who invoke the industry and Silicon Valley to illustrate their views of the causes of economic growth there has always been another powerful actor behind development:

> Far from being created by ineffable scientific genius and heroic entrepreneurialism, the electronics industry in its infancy proceeded ultimately from extensive state intervention. (Gordon and Kimball, 1986)[17]

Also running counter to the image of the industry as a saviour of depressed economies is the much-discussed possibility of large-scale technological unemployment as a consequence of the rise of microelectronics and even within the industry there have been many well-publicized slumps, redundancies and company failures. Many have responded to the point about technological unemployment by accepting it with the qualification that the prospects of job losses would be worse for those who do not embrace high tech. But while this is probably true it does not follow from this that embracing high tech, boosting sunrise industry, is the solution to problems of recession and unemployment. This false argument is so common that we have christened it 'the new technology non-sequitur'. Yet along with the rest of the ideology of high tech it is part of the rhetoric of 'boosterists', as they are known in America. As such people pin their hopes on cornering some of the sunrise industries, it is not in their interests to acknowledge such problems for to do so might undermine what they hope will be a self-fulfilling prophecy.

This gives a clue to a further aspect of the ideology of high tech – its significance in the context of the recession – for it is this that creates the need for good news and which throws the handful of growth industries into such sharp relief. In Britain, where this need is greater than in other advanced industrial countries, it accounts for the rather desperate lionization of figures such as Sir Clive Sinclair, the entrepreneur scientist responsible for the first 'bargain basement' home computer.[18]

As we have already implied, Silicon Valley has a special place (literally) in the ideology of high tech, firstly as it is commonly misrepresented and secondly in its assigned role as a generalizable phenomenon. The Valley consists of more than millionaires, whizz-kids, R&D, entrepreneurialism, Porsches and jacuzzis. It also contains large numbers of low-paid, chiefly Hispanic and female production workers, often using hazardous materials and unable to afford to live near their place of work because of rocketing land values (Bernstein et al., Saxenian, 1984). Secondly, and of fundamental importance to the arguments of this book, the idea that other places can hope to clone Silicon Valley's development is deeply misconceived. This is because it is an integral part of a larger structure of combined and uneven development – the international spatial division of labour in electronics – and not an island of growth, divorced from developments elsewhere. Within this structure, different places have different yet interdependent roles, so that changes in one part imply changes in another. Since there is only a limited amount of prestige functions within the industry to go round, the dreams of the hundreds of local boosterists wanting to emulate Silicon Valley could only be realized simultaneously by a miracle equivalent to that depicted in the parable of the feeding of the thousands with two fishes and five loaves. So given that firms – often the same ones – perform different functions in different parts of the world, be it Taiwan, Mexico or

Scotland, it is absurd to consider such locations as further Silicon Valleys in the making, though some may at least move 'up-market'.

The problem here is one which logicians term 'the fallacy of composition', namely, the assumption that what is possible for an individual (Silicon Valley) must therefore be simultaneously possible for all individuals (anywhere wanting such an industrial agglomeration). Yet the limits of markets and competition together with interdependencies between different and unequal elements of the structure prevent anything more than very limited replication. And it is this fallacy, in the context of the desperate search for good news, that lies behind the farcical or tragi-comic prefixing of the names of places having minor (and qualitatively different) concentrations of electronics, such as 'Silicon Glen' (central Scotland) and 'Silicon Fen' (the Cambridge region).

So while we do not underestimate the significance of the term high tech, we intend to resist the content of the ideology; the electronics industry is a capitalist industry no less than any other and it is therefore subject to the same pressures as any other. Nor, however, can it usefully be treated as a black box or as identical in every respect to other capitalist industries; to understand the nature and possibilites of electronics it is necessary to go deeply into its special characteristics while linking these to its general properties as a capitalist industry.[19]

PART II

The Electronics Industry and its Global Structure

In part I we argued that, as a highly internationalized industry, the development of electronics in different parts of the world could only be understood in a global context. It is to this international level that we now turn. The following chapters are intended first to introduce the industry and second to explain its changing pattern of uneven development at the international level. The first task is necessary for the second, given our earlier comments on the mediations of capital accumulation, for we believe that the mechanisms of competition and corporate behaviour are far richer than is recognized in most theoretical discussions.

While mediations such as technology and the social forms of capital, government intervention and the nature of national economies and markets all interact simultaneously, it is of course impossible to explain them simultaneously. Therefore, whenever we refer to a particular mediation, such as the institutional forms of capital, it is essential to keep in mind that this has not arisen in isolation but through interaction with other mediations and, as we hope to show, the global spatial organization of the industry has not been merely an incidental effect of its development but an important constitutive moment.

We have chosen to split the industry up into four main sectors – semiconductors, consumer electronics, computer systems and telecommunications. While these do not totally exhaust the electronics industry they dominate it and, although there are many interactions between these sectors, each also has a certain unity and distinctiveness which imparts different forms to their uneven development.

As one quickly discovers when specializing in a particular industry, it is easy to get pulled into progressively more detailed treatments of the characteristics of technologies, companies and government policies. Our criterion for deciding how much detail to include is that it should make some significant difference to the processes and patterns of uneven development at the scales in which we are interested. This may not be sufficient for the industry specialist, but any finer detail is in our view likely to be of limited and transitory interest. Certain themes – particularly regarding the influence of national economies – run through all four sectors. To avoid repetition we will discuss these most fully in the first of our sector chapters, on semiconductors, mentioning them more briefly in the remaining chapters and only enlarging where the sector in question is affected differently by them.

3

Semiconductors

The components sector has been the most dynamic part of the electronics industry and its technological advances – particularly the chip – have affected and will continue to affect not only the other electronics sectors but virtually every other industry. Within components, it is semiconductors which have provided the heartland technology of the microelectronics revolution. Although this subsector employs only about 300,000 world-wide, its strategic significance is enormous as is recognized in the fact that most advanced industrial countries and some developing countries have a support programme for microelectronics in general and semiconductors in particular. And, since this sector has become highly internationalized, the possibilities for such programmes are strongly influenced by the global pattern of development.

Some key features of the semiconductor industry

The semiconductor industry began about 37 years ago with the development of valve technology and since that time has seen a remarkable rate of technological progress, the three outstanding innovations being the transistor (1957), the integrated circuit or IC (1961) and the microprocessor (c.1971). The development of the IC made it possible to combine the functions of large numbers of formerly discrete components, such as transistors, on a single chip. Large scale integration (LSI) combined thousands of components on a chip and now, with the development of very (or ultra) large scale integration (VLSI or ULSI), chips can now be made which perform a million functions. While there are many types of chips, we shall pick out for special attention two major families: microprocessors (which perform the functions of a computer's central processing unit) and memories. We shall also have occasion to refer to an additional minor, but rapidly growing, group called application specific ICs (ASICs) which are customized wholly or partly to users' particular needs. The significance of these types of semiconductor for uneven development will become clear shortly but it derives from the fact that the former are the most demanding in terms of complexity and product design while the latter are the most demanding in terms of production process technology and organization.

The industry is noted for its extraordinary market development. Against the background of the world recession, demand has grown at an average of 30 per cent per annum, though employment growth has been modest by comparison. The rapidity of this growth is attributable to the dynamics of the relationships between markets and the use-value and exchange-value characteristics of the product. As the miniaturization – or integration – of chips has proceeded, the cost of electronic circuitry, and more specifically of computing, has fallen dramatically, while their performance has improved equally dramatically. Over the past 30 years, hardware performance per unit of cost has advanced by a factor of nearly one million. The impact of this development on user sectors, such as telecommunications and computers, has made the semiconductor industry a truly propulsive one.

As might be expected, these developments have radically altered both the kinds of know-how needed and the pattern of value-added in electronics and other user industries. One result of this is that the social organization of the electronics industry has changed as intersectoral boundaries have shifted. The ever-increasing miniaturization of circuits has meant that chips have become progressively less like components and more like systems, thereby encouraging

Table 3.1   Worldwide semiconductor sales

	$ billion	percentage change
1982	14.1	
1983	17.8	+26.3
1984	26.3	+46.0
1985[a]	21.6	−17.0
1986[a]	25.5	+18.0
1987[a]	31.3	+23.0

[a] Estimate.

Source: Semiconductor Industry Association

a convergence between the components industry and the systems producers in computers, telecommunications and in military and industrial equipment.

Despite the rapidity of its growth the industry is exceptionally cyclical, with 'chip gluts' succeeding 'chip famines'; in its short life there have already been four slumps, in 1970–1, 1974–5, 1981–2 and 1985–7. The amplitude of these cycles has also widened, the last boom being the biggest (43 per cent increase in sales between 1983 and 1984), and the last recession being the first to record an absolute fall in demand (−17 per cent for 1984–5) (table 3.1).

The cycles can be explained by reference to the cost and price structure of the industry. Such is the rate of technological change that, if companies are to stay in business, they must introduce new generations of more powerful products at regular intervals (e.g. every three to four years) and for each new range R&D and capital costs, and with them optimal scale economies, increase steeply. For example, the 16k memory chip (c.1981), which succeeded the 4k chip of 1977, required five times as much capital investment as its predecessor, and this ratchet effect has continued. So, while the latest chip plants can cost as much as a small steel mill, their equipment is likely to be technically obsolete in two to five years. Consequently, firms often spend a quarter of their sales revenue on new plant and equipment (Ernst, 1981). Once volume production of a new chip begins costs plummet and firms (especially the Japanese) pursue aggressive pricing policies to keep out competitors who are on the same learning curve. Thus the 256k memory chip has fallen in price from $110, at the time of its introduction, to $3.75 in little over a year. Competition is further increased by the fact that the improvements in chip performance allow demand to be met by fewer chips (Duncan, 1981; Rada, 1982).

Not surprisingly, the risks of operating in this market are considerable, though so are the gains from being a year ahead of competitors. It is these gains, plus the strategic significance of semiconductors, which lead firms to commit such resources to the industry. However, they are also responsible for inducing periodic crises of overproduction as each firm hopes that it will be the one to be first on the market with a new generation of products. When overproduction leads to margins being cut, investment in the next generation of products often has to be postponed with the result that, when demand catches up with supply again, capacity is stretched, thereby triggering another flurry of investment in new capacity. This classic type of overproduction crisis is further exacerbated by chronic over-ordering during the boom periods by purchasers anxious to safeguard their supplies in a tight market. But the most recent slump (1985–7) was far deeper than any earlier one because of a recession in the computer industry itself, particularly through the shake-out of small firms making personal computers which provided much of the demand for standard chips.

Despite the huge investment costs which present formidable entry barriers to the industry, many firms have tried to enter or re-enter the standard IC market, often with large state subsidies. For example, Philips and Siemens, the two largest European firms, have received $400m in subsidies for their attempt to enter the 1 megabit and 4 megabit memory markets, but then Texas Instruments, the largest US merchant producer, is reported to have invested $1.3b. While many of these attempts have been underfunded, it nevertheless intensifies competition and the tendency towards periodic overproduction. It is a measure of the perceived strategic significance of the industry that until recently these conditions do not seem to have deterred would-be entrants and their state supporters.

The firms which were hardest hit in the 1985–7 recession were the big American merchant producers. By comparison, captive producers such as IBM and AT&T have suffered less and consequently have gained market share, presumably because of their access to cheap finance, their greater control of semiconductor demand and their ability to exploit synergies between component and system design (Ernst, 1986). The present position of the Japanese producers is intermediate in the sense that they are part captive, part merchant and so can take advantage of inter-sectoral synergies, although they are more exposed to market forces than pure captive producers.

In response to the recent slump, many firms which are already heavily committed to mass-produced chips are trying to diversify, usually downstream into 'systems' for which value-added and markets are more secure. Another increasingly popular strategy being taken up by both established companies and would-be entrants is in ASICs, for which demand is expected to grow rapidly. However, even here many of the ventures have been underfunded and overcapacity is already imminent, showing that overcapacity can occur even in niche markets (Rada, 1985; *Fortune*, 6 January 1986).

As with any industry, there are many labour processes but the most common in IC production can be broken down into the following four stages:

1 R&D, design, marketing and other pre-production managerial activities;
2 wafer fabrication, a capital-intensive process involving etching microcircuits onto thin discs of silicon called wafers;

Table 3.2   Shares in world integrated
circuit production

	1978	1981	1985
US IC total	68.2	71.6	66.8
Merchant	48.2	48.4	43.8
Captive	20.0	23.2	23.0
Western Europe	6.7	6.3	5.1
Japan	17.8	20.7	26.7
Rest of world	7.3	1.4	1.4

Note: Different sources have published widely
varying statistics. Some, for instance, put the
Japanese share in 1985 as high as 41 per cent.

3 assembly, i.e. dicing the wafers into individual chips, soldering
   wires to them and mounting them on carriers;
4 final testing – more capital- and skill-intensive.

As we shall see, the different characteristics of these stages of
production have important locational implications.

The global development of the industry

Although the sector is dominated by large multinational companies
and is thus highly internationalized, the global pattern has been
strongly influenced by the character of the multinationals' home
markets and by state support and protectionism. As we shall see,
this national character does not merely consist in firms passively
absorbing or responding to national characteristics but in their active
participation in national alliances as a weapon in competition.

The United States

From soon after its inception the industry was dominated by American
firms. Three factors stand out in explaining this lead. The first
was the USA's strong position in scientific research on electronics
after the war; in addition to leading research programmes at
universities such as MIT, Berkeley and Stanford, the fundamental
research carried out at Bell Laboratories was a prolific source of
inventions. Moreover, interaction between the scientific establishment
and industry was strong, particularly as Bell was obliged to license
innovations to other firms at low cost, thereby facilitating diffusion.

The second factor was the enormous size and sophistication of the national market. However, as we noted in chapter 2, markets are not simply given but have to be created. In this case the principal formative role was played by Federal expenditure during the fifties and sixties on military and space applications, particularly the Minuteman missile and the Apollo space programme, in which the semiconductor technologies were chiefly developed. This expenditure was far in excess of that of other countries, and constituted a formidable advantage. So contrary to the 'strong market–weak state' stereotype, the US Federal government, via the Department of Defence (DoD) and NASA, historically played a critical role in the development of the country's electronics industry. The DoD sustained early supply by being the principal purchaser and by funding both R&D *and* production capacity; in 1962 the US government accounted for 100 per cent of IC sales in the USA (Nelson, 1982; McKinsey & Co.,1983). This provided a vital base load for the firms to augment their production capacity and, later, to diversify into commercial markets. Where government procurement in European countries created only small niches which did not induce volume production, the scale and duration of the US projects were sufficient to encourage a volume production capability as well as technical virtuosity.

Thirdly, in contrast with Europe, the American producers had the advantage of a pattern of demand which came to be dominated by the computer industry (which itself led the world) and which required both state of the art design-intensive chips and large volumes of standard chips. This kind of demand exposed suppliers to fierce price competition and hence process innovation and cost-cutting (Borrus, 1985; Henderson, 1986). As we shall see, this synergy between the semiconductor and computer industries accounts for America's near monopoly of key innovations in these fields (see Soete and Dosi, 1983). In recent years, however, the influence of the DoD has increased again with rising military spending, especially connected with the Strategic Defence Initiative ('Star Wars').

Traditionally, the American semiconductor industry had been concentrated in the north-east of the country, in a belt from Massachusetts to Maryland, but such were the changes involved in developing ICs as regards expertise, personnel and business organization that this area ceased to hold locational advantages for the new products and the nascent IC industry was 'captured' largely by California, particularly the area at the southern end of San Francisco Bay which later came to be known as Silicon Valley (Scott and Angel, 1986). Here, under the influence of local aerospace and military research labs and markets, a core of semiconductor firms emerged. At the same time a unique but broadly based concentration of scientific and managerial expertise developed; though widely cited, the rise of dynamic electronics departments at Stanford and Berkeley universities appear to have been a part of this process rather than an initial cause (Saxenian, 1985). These developments have helped to localize subsequent growth in the US industry in the area so that now, with over 270,000 jobs in microelectronics, it is probably the world's largest single concentration of such industry.[1] As Scott and Angel show, in addition to HQ and R&D functions, small batch custom chip production has also concentrated here, attracted by – and in turn attracting – the agglomeration of suppliers, subcontractors and specialist customers in the area. A further cause and effect of this localization of production is the plentiful supply of cheap un-unionized labour, mainly Hispanic and female, in California; contrary to the image of the area, semi-skilled operatives are the largest single group of workers in the valley (Bernstein et al., 1977). Meanwhile mass production of standardized chips has tended to go to cheaper areas both inside and outside the USA, as their production is more routinized and needs little interaction with suppliers and buyers. This tendency has been accelerated in recent years by the overheating of the land and labour markets in Silicon Valley. At the same time, the north-east has remained dominant in discrete semiconductors.

Contrary to the Eldorado image of Silicon Valley, the present slump has brought shutdowns, layoffs and pay cuts, even in prestigious firms like Intel, National Semiconductor and AT&T. Between December 1984 and October 1985 employment in the American industry fell by 54,000 or 19 per cent and the merchant firms lost over $500m. Remarkably, all but one major US merchant firm has withdrawn from the most competitive volume markets (for dynamic random access memories) as capacity utilization fell to 30–40 per cent (de Jonquieres, 1985e; Ernst, 1986; Office of Technology Assessment, 1986).

Western Europe and the American invasion

Until the mid to late 1960s, American semiconductor firms such as Texas Instruments either merely exported to Europe or used their European subsidiaries only for making mature products, reserving new state of the art technologies for the more advanced US markets. In this role, they did not challenge the major European firms which had formed a stable electrical–mechanical oligopoly (see Teulings, 1984). Having been at least the equal of the US industry prior to the transistor, the European companies were able to maintain their position by imitating and licensing the more advanced American products. The break-up of this situation began with the introduction of ICs by existing and new American firms. The more cumulative nature of the learning process in ICs suddenly widened the technology gap, and hence the imitation lag in Europe. The US firms then began to produce and sell the new products much earlier in Europe, thereby cutting their imitators out of the advanced markets.

The situation was exacerbated by the fact that many ICs replaced, rather than supplemented, the discrete components in which the European companies had considerable strength – a classic example of the power of product innovation as a competitive strategy. Then, in 1971, in response to a slump in demand in the USA, the American firms started a price war in Europe, though not in the USA, which further advanced their position (Dosi, 1982). Also, whereas the European firms were inhibited by their inheritance of a fragmented production base with each country being supplied by a separate plant producing according to national technical standards and marketing constraints, the American firms, with the new products, were able to establish pan-European technical standards and use a new set of plants geared to producing for several countries (*Financial Times*, 26 October 1982). The result is that now American-owned firms account for between a quarter and a half of European output. In other words, the small size of the European national markets (totalling only 16 per cent of the world market) was itself a disadvantage, though more to the indigenous firms than the US-owned firms.[2]

The influx of American firms such as National Semiconductor, Motorola and Hughes into Europe in the seventies and eighties included not just point-of-sale and assembly and test functions but their only significant offshore investments in wafer fabrication plants, many of them located in central Scotland. The reasons for this recourse to foreign direct investment were to get behind the 17 per cent tariff barrier on semiconductor imports into the EEC, to liaise with local military customers needing more specialized chips and to take advantage of cheap technical labour, in the case of Scotland costing only 40 per cent of that in California. As so often happens, some of these European plants have now come to play a less parochial role and produce a number of standard products for global distribution (Henderson, 1986).

The reaction of most of the European firms was to withdraw from the fiercely competitive mass markets for standardized chips into the more specialized markets and in-house production, chiefly for defence uses where they were, to a certain extent, insulated from the law of value by the special pricing conditions attaching to government defence procurement. For example, GEC of Britain withdrew from the standard IC market in 1971 at the same time as Intel was beginning to market its revolutionary microprocessor. While specialist products might be technically sophisticated the processes by which they were made did not have to be; indeed their production required wholly different types of organization from that for low-cost mass production and marketing. Government telecommunications contracts also provided a protected refuge (e.g. for Plessey and GEC in Britain, and until recently SGS in Italy), but the economic effects from this state involvement were quite different from those in America. However, even in the telecommunications market, technical superiority and European bases gave American firms such as ITT limited entry into these markets too. In the end, Philips (Netherlands) and Siemens (FRG) were the only European firms large enough to approach the US firms' R&D expenditure and hence retain a competitive toe-hold in the markets for standardized ICs.

Technical complementarities (or their absence) also affected the changing relative strengths of capital in the two continents. Electronic components for military uses tend to be different from those for civilian products, adding to the difficulties facing firms in changing from defence to consumer work (Maddock, 1983). This is particularly the case in Europe where demand was limited and where government support tended to be for research only and not production. Most European applications incorporating ICs are demanded in runs of less than 10,000, and many in less than 1,000, and as we have already indicated the crucial computer market is much smaller and less advanced than in the USA. Even where European semiconductor firms have come up with highly innovative products for commercial markets, they have encountered difficulty in selling them.

This illustrates how technological interdependencies lend their own influence to the temporal and spatial dimensions of uneven development. They imply that any country which lacks a capability in semiconductors will become technologically dependent across its electronics industry as a whole. Buying in standard chips from merchant producers may be cheaper in the short run and less risky, given the economics of chip production, but it entails high software costs (for adapting standard components to specialized functions) and means foregoing long-term advantages in designing special-purpose chips in-house. It therefore implies reactive rather than pro-active competitive strategies – in other words, greater vulnerability. And, as we have already noted in the case of ICs, the effects of being distant from the 'technological frontiers' (Dosi, 1982) are dramatic for the process of integration actually swallows up markets for older discrete products.

The 'runaway' phenomenon

It was during the build-up of IC markets that some of the best-known chip firms further developed their international spatial division of labour in a different direction – towards the Third World, in search of cheaper and more exploitable labour. In part this corresponds to the pattern associated with the *New International Division of Labour* thesis and is often cited as a prime example of it (Fröbel et al., 1979). For instance, many major US chip firms have the following sorts of corporate spatial divisions of labour.

1 R&D, design and other pre-production activities needing skilled scientific and managerial labour, plus small batch specialist production, are located in the USA, mostly in Silicon Valley.
2 Wafer fabrication requires highly skilled process engineers as well as semi-skilled operators and is therefore likely to be located in a rich country, e.g. in the USA itself or in Europe.
3 Assembly is a more labour-intensive stage of production and requires semi-skilled labour. Given the portability of the product, transport costs constitute less than 1 per cent of total costs.[3] Third World locations have become popular, at least for standard chips, though not for specials which are produced in volumes too low to make this worthwhile and for which assembly costs are minor in relative terms. Assembly labour costs may be as little as 4 per cent for the former and as high as 33 per cent for the latter (quoted in Dicken, 1986). Typically the low-volume work is done in-house together with wafer fabrication or subcontracted out locally.
4 Final testing is more capital- and skill-intensive and tends to have been carried out in the USA or in the newly industrializing countries where there are sufficient local skills.

This corporate international division of labour is invariably structured on gender lines, with repetitive, clerical and manual work being carried out almost exclusively by young women, while work defined (rightly or wrongly) as more skilled is the preserve of men of a variety of ages (Wong, 1985).

The locational attributes of points 1, 2 and 4 are generally true of the industry as a whole and apply, with appropriate adjustments for home base, to European and Japanese firms. However, it is point 3 – the 'runaway industries' – that have aroused most attention, especially on the Left. In 1962, Fairchild, a leading US semiconductor firm, established a labour-intensive assembly plant in Hong Kong. It was followed by a procession of rival firms from the USA, locating first in what we now call the newly industrializing countries and

Table 3.3   Principal developing country
locations of semiconductor
assembly

Country	Value of OECD imports of semiconductors ($ million)
Malaysia	1066.4
Singapore	820.6
Philippines	482.1
Republic of Korea	385.1
Taiwan	359.4
Hong Kong	202.7
Mexico	145.2
Thailand	85.5
Indonesia	62.5
El Salvador	49.3
Brazil	30.7
Barbados	14.0
Total	3703.5

Source: United Nations Center on Transnational
Corporations, 1983, p. 35

later diffusing out to countries with still cheaper labour, such as the
Philippines (Chang, 1971; Ernst, 1981; Rada, 1982). A smaller number
of plants was set up in Latin America, particularly in Mexico (table
3.3). In fact, the attractiveness of such locations to capital lay not just
in their low wages, which were often between a tenth and a quarter
of American rates. Also important are longer working hours, greater
scope for multiple shifts, replaceability of labour, state repression of
labour organization, tax holidays and free repatriation of profits.
Typically, young women would be recruited for assembly work, often
from rural areas and having no previous experience of industrial
work. It is not merely their origins from poor areas which made them
both cheap to employ and highly productive but their subordinate and
vulnerable position within these societies. Nevertheless, although such
workers have frequently been described as 'docile' by both companies
and governments, there have been many strikes, sometimes at considerable risk to individuals. Much has been written on the Left about
these matters and on the oppressive conditions regarding health and
safety, pay and suppression of resistance in these factories.[4] However,
while it is important that these conditions be exposed, there is a
widespread tendency on the Left to exaggerate their relative quantita-
tive significance in global patterns of investment within the industry.

The picture is also more complicated and fluid than that suggested by the *New International Division of Labour* theorists. First, the Third World locations are favoured not by all but by a particular subset of semiconductor firms, namely American merchant producers of standardized chips such as National Semiconductor. The captive producers in the USA (i.e. those who produce for in-house use), which in the case of IBM and AT&T rank in output with the biggest merchant producers, have until recently kept assembly onshore (Scott, 1986a). This is partly because of their different priorities and cost structures but is also due to their greater use of automated assembly.

Second, the predominance of American firms following the runaway strategy reflects not just their dominance of the sector but the fact that the US tariff structure is much more favourable to the (re)importation of semiconductors from offshore plants than that for Europe. While the latter has a 17 per cent tariff, American importers only pay duty on value added abroad (see Organization for Economic Cooperation and Development, 1985). However, this does of course give American firms a motive for addressing the European market from European rather than Third World bases, and this has been a contributory factor in the establishment of assembly and wafer fabrication plants in the EEC, especially in Ireland and Scotland. Third, as chip integration becomes greater, the relative amount of assembly work to be done becomes smaller, thereby shifting the burden of costs towards water fabrication.

Fourth, in recent years, many of the merchant producers have been turning to automated assembly. With automation, one worker with two weeks' training can replace 30 manual assemblers with three months' training each (Posa, 1981). These changes may have slightly reduced the attractiveness of Third World locations relative to the economies of co-locating automated assembly with wafer fabrication in developed countries. This also has the further advantage of reducing the large in-process inventories which are needed to maintain continuity of production between plants on different sides of the globe, thereby allowing more rapid feedback to wafer fabrication and hence better quality control. Given that minimizing the reject rate is probably *the* key variable as regards cost competitiveness in IC production, co-location may be worthwhile even though labour costs are higher.[5] However, only a few automated assembly plants were located onshore and some of these, such as Motorola's new plant in Phoenix, Arizona, have been abandoned, largely due to the recent chip slump.[6] In any case, automation can be introduced offshore and there is no evidence of Third World plants being abandoned; moreover, a few new firms such as the revamped SGS-Ates of Italy are making use of them.

Finally, the runaway industry phenomenon has been modified by rising wages in some of the early Thirld World locations, which have reduced their attractiveness for labour-intensive low-skill work. In particular, the differences between wage levels in the USA and the newly industrializing countries, especially Hong Kong, have narrowed. This is of course a reflection of their success. Not just any low-wage country is able to take over their former role, however, for infrastructure and political stability must be assured.[7]

None of these influences operates singly; rather they enter into trade-offs and any specific change rarely does more than tilt the balance slightly. Given the many varieties of chip and their different cost structures, it is hardly surprising that the locations of assembly plants differ widely. Moreover, the practice of large firms of spreading their risks and hence avoiding concentrating all their investments in a single optimal location also contributes to the dispersion of the industry.

Japan

The Japanese story is different again. Here the electronics industry received no stimulus from military spending in the American fashion, nor any distraction from it, as happened in Britain. Although the home market is large it was initially technically unsophisticated. However, by a vigorous process of imitation and improvement behind national barriers to inward investment, licensing and, until 1974, to imports of semiconductors, it has become a major force in the world market, overtaking Europe to become the second largest producer, accounting for over a quarter of the world market and assuming the lead in some products. A number of reasons can be put forward for this success, many of them applying to other sectors and industries in Japan.

Many observers have commented on the role of the Ministry of International Trade and Industry (MITI) in directing the development of this and other industries, though as we shall see the government has also influenced industry via other routes. Collaboration with MITI enabled the Japanese firms to forge a coherent long-term strategy, targetting particular products and developing new complementarities between components, consumer electronics and computers and helping to determine common technical standards for products. Unlike the USA, where innovations in ICs were introduced first by new specialist firms, IC production in Japan was assumed by large, established, vertically integrated firms like Hitachi and Toshiba which were able to cultivate the synergies between components and systems production. Being 'semi-captive' producers of semiconductors they have been able to cross-subsidize this branch from their other activities; as the American merchant producers complain, the Japanese companies have 'deep pockets' and do not regard their semiconductor divisions as profit centres. MITI has also arranged low-cost finance, export incentives and tax credits for cooperative R&D and investment; between 1976 and 1982, it gave £500m in grants and loans to semiconductor companies and £1.5b for related sectors such as computers. Above all, MITI has facilitated the transfer of technology within and between firms. Basic technological advances are considered common property but while there is often a division of labour between firms, or blocks of firms, regarding types of product area or technology actual product developments are not shared. Most unusually, despite heavy protection by tariff barriers, there has been vigorous competition within the domestic economy. Protectionism has not therefore had the kind of damaging 'feather-bedding' effect on domestic industry that it has had elsewhere. Formal protectionism against imports and inward investment has now officially ended but it continues unofficially because of tight customs regulations, reluctance to buy foreign goods and the difficulty of usurping the major firms' traditional sources of supply, given the tight and enduring links between large firms and their suppliers. These bonds are common in Japan even where ownership links are absent, i.e. without formal vertical integration.

The special relationship between Japanese industrial and financial capital has also made a significant difference to its international competitiveness. Government support for priority sectors means that their firms have access to low-interest loans.[8] They also have little recourse to equity and have higher debt–capital ratios than American or European firms, e.g. 60–70 per cent as against 16–18 per cent in the USA (Rada, 1982). The latter features owe much to the firms' membership of industrial groups. These always include one or more banks and operate on an 'umbrella' principle in which members cooperate and are closely involved in each other's long-term future. As Dore (1985) puts it, most investors are knowledgeable committed investors, not interested in short-run speculative financial gains. The contrast with American, and especially British, capital could hardly be stronger.

This means that the Japanese firms can continue to operate and invest with lower rates of return than the Americans – an enormous advantage in the battle for world markets, particularly in semiconductors. Repeatedly, while the US firms, especially the merchant producers, have had to postpone or cancel new product development during the slumps, the Japanese have been able to keep up investment and hence be first in the market with new products and new production capacity when the ensuing boom arrives. For example, while the

US firm National Semiconductor has had to put the development of its 256k memory chip on ice, some Japanese firms have already had these products on the market for a year and are announcing the introduction of the next generation of chip, the 1 megabit memory. The advantage of the Japanese firms in the booms was evident in the last 'chip famine'; when IBM could not meet its demand for memory chips internally, it turned to Japanese suppliers as the US merchant producers could not meet the demand in sufficient quantity or quality. This advantage, coupled with extremely aggressive pricing (again facilitated by favourable financial conditions), has given Japan overwhelming dominance in memories, as is clear for market share figures for successive generations of chip: over two-thirds for 64k; 98 per cent for 256k and probably the same for 1 megabit chips.

These operating and investment cost advantages allow Japanese firms to adopt a much longer-term perspective on innovation and product development than in the West, a characteristic which in any case is supported by MITI. This effectively protects new products from premature death by excessive exposure to market forces and allows superior preparation and targetting of new markets. Nevertheless, the recent falls in demand from the computer sector have brought large drops in sales to America, and falls in employment in Japanese companies in the region of 5.5 per cent, showing that even the most efficient producers are vulnerable in a slump (Ernst, 1986).[9]

Many other features of Japanese industry have been cited as explanations of its astonishing rise in this and other sectors. This is a complex and poorly understood subject in the West and we can do no more than provide a brief explanatory sketch here. Traditionally, the Japanese advantage was explained in terms of low wages but pay levels have now virtually caught up with leading Western economies, at least in the large firms though wages in small subcontract firms are often 40 per cent lower. But, in any case, wage costs do not explain productivity and quality advantages. These features are often explained by the defeat of militant trade unions in the private sector in the early fifties and their replacement by pliant company unions, often virtually extensions of company personnel departments.[10]

But pliant labour only provides an enabling factor in the development of productivity and quality advantages. What is crucial is the actual material organization of production processes and their research and administrative back-up. 'Total quality control' is crucial in this, involving much greater efforts than in the West to eliminate defects and waste in both bought-in parts and in-house production, where Western companies have often tolerated them in the pursuit of speed and volume. This is particularly important in chip production where the intricacy of operations means that the most critical cost variable is the proportion of defective chips. So the renowned obsession of the Japanese with quality is not public relations but an effective way of reducing costs by eliminating waste. This is helped by the use of quality circles, the training of workers in multiple rather than single skills and by insistence upon flexibility, together with the knowledgeable and persistent involvement of process engineers and technically competent managers in the piecemeal improvement of production. This makes for not only higher labour productivity but higher utilization and turnover of (constant) capital and much lower inventories than in the West: in short, a much leaner form of organization. Flexibility in restructuring is also assisted by the practice of rotating managers and top workers between different parts of firms so that internal coordination is improved, and this in turn is supported by lifetime employment for core employees (invariably male).[11] Finally, restructuring and product innovation are facilitated by enduring and intimate relations with suppliers. Even where these are separately owned, they coordinate product development with the plans of major users. For example, Japanese semiconductor producers have benefited from their close links with semiconductor production equipment manufacturers in improving their process technology.

Historically, after an initial success with transistors in the 1960s, considerable long-term effort was devoted to catching up with the USA. Besides developing their advantages in the organization of labour, Japanese semiconductor producers have developed automation more fully than their Western competitors. The elimination of workers from chip fabrication and assembly has been given special priority because it is often dust from workers' bodies which contaminates the production process and leads to defective products. More generally, the ability of Japanese firms to capitalize on learning-by-doing made their use of automation more effective (i.e. it raised capital productivity higher) than in the West. The result of these characteristics is that, even though product technology in semiconductors still lags behind the USA in many fields (e.g. microprocessors), their production process organization is unsurpassed.

From the outset, a particular and well-chosen range of chips was targetted – memories – for it was these in which the Japanese firms could take most advantage of their lead in high-quality volume production. For these products,

> ... stringent quality control in the production process was relatively more critical than such factors as innovative design in ensuring product competitiveness. Circuit design is a less formidable task in the case of memory chips than in the case of microprocessors. Moreover, the

former devices involve no software development and service [in which the Japanese are weak] as do the latter. To a far greater extent, successful competition [in memories] depends on the refinement of mass production techniques and process know-how. In memories high yields [i.e. low rejects] – hence tight process control – are the crucial variable since they make possible the dramatic price reductions which have come to characterize competition in this segment of the i.c. market. (United Nations Center on Transnational Corporations, 1983, p. 275)

An additional consideration in this specialization on standard chips within the sector has been the structure of demand, which has been dominated by Japan's innovative consumer electronics industry, taking 57 per cent of output, as against 21 per cent in the USA (Borrus, 1985). However, the Japanese firms are not resting content with this position but are gaining competence in more advanced products such as microprocessors and they have already established dominance in the more basic ASICs, with Fujitsu the clear world leader (Ernst, 1986). At the same time, a growing proportion of demand is coming from more advanced users in computing and telecommunications who require more design-intensive chips. As we shall see, this strategy of playing to their competitive advantage in mass production first and then moving up-market is also apparent in other sectors. As Dosi (1982) comments, where other countries took their place in the international division of labour as given and hence took licensing (i.e. the institutionalization of technological dependence) and inward investment as parameters, Japan treated them as variables and strove successfully to change them.

Japan's competitive advantage has thus become the USA's disadvantage. The loss of market share in this, the main 'cash cow' of the American merchant firms, has squeezed the revenue available for diversification into other kinds of product (in a way, this is reminiscent of the dilemmas facing Third World countries which are overly dependent on a single export product). But, perhaps more seriously, they are in danger of losing an important means by which they can retain proficiency in volume production process engineering.

Recent foreign direct investment in advanced countries

The conditions in Japan regarding manufacturing also explain much about Japanese companies' strategies regarding internationalization. In addition to government restrictions on the export of capital, the Japanese firms have been deterred from investing abroad by the difficulty of reproducing overseas the advantages they enjoy within Japan. In the case of Third World production bases, Japanese producers have for many years largely sacrificed the once-and-for-all advantages of cheap labour etc. and concentrated on the longer-term and progressive advantages of automation. Its success in this strategy has increased pressure on American and European rivals to automate more. Consequently, Japanese firms have made much less use of cheap Far East labour locations than the Americans have, and where they have had to use them for cost reasons they have also used more automation. According to Rada (1982), Japan's offshore semiconductor production amounted to only 3 per cent of its total domestic production while the equivalent figure for the USA was 37 per cent of total domestic production. This situation is mirrored in electronics generally; whereas US and EEC imports from Korea, Taiwan and Malaysia totalled $6,028m and $1,528m respectively in 1983, imports for Japan equalled only $590m (Organization for Economic Cooperation and Development 1984). However, a combination of import restrictions, the rising value of the yen (itself of course an indicator of Japan's success) and the extreme price competition of the present slump is forcing more Japanese firms to manufacture in south-east Asia (Ernst, 1986).

Japanese investment in semiconductors in other developed countries began in the late seventies, the overriding reason being to pre-empt protectionism. Secondary motives have been to improve access to sophisticated markets as they move up into products requiring more involvement with buyers, to establish a higher profile and greater credibility with major industrial customers and to monitor US technology more quickly. NEC, Oki, Hitachi, Toshiba, Fujitsu and Mitsubishi have all invested or are about to invest in fabrication and assembly in the USA.

European manufacturers have also invested in the USA, obviously not for reasons of cost but for access to the US market (50 per cent of the world market for semiconductors). The needs to catch up with American technology – a 'bootstrapping' strategy – and to establish domestic producer status have been prime motives behind European producers' investments in the USA. However, this can entail making products which are too advanced for the European market. For example, the British chip firm Inmos sold 80 per cent of its output in the USA and its chief executive commented that '. . . many UK customers complain that its products are too advanced for their needs. Inmos is an irrelevance to European industry at this stage' (quoted in de Jonquieres, 1985b).[12] In fact, despite the fact that Inmos was started, with heavy state backing, as a reply to foreign dominance in chips, over half its investment and employment has been located in the USA, the primary reason being the market and a secondary one the greater availability of leading skills in silicon technology. There have been many other similar cases where such moves have been considered essential for success.

Several European companies have made deals with American firms, and, as has often happened within the American industry itself, large diversified electronics firms lagging behind in ICs have taken over or bought into small innovative US firms, exchanging cash for know-how and market share. For example, Siemens owns 20 per cent of Advanced Micro Devices, while Philips has bought Signetics.[13] Thanks to low volumes of output and the EEC tariff, European investment in the Far East has been small.

Many firms from each of the core countries are investing in R&D and design centres in other advanced countries, in order to tap local software skills and provide facilities for customers to design ASICs. Even standard products such as mass-produced microprocessors require some interaction with customers to help them define their needs and learn how to use the products. In other words, the location of investment by the semiconductor multinationals is heavily influenced by the need for access to advanced markets in the shape of computer or telecommunications firms and the like, as well as by the distribution of scarce relatively immobile technical skills – skills which are also much cheaper in countries such as Britain than in the USA. The value attached by European governments to acquiring a capacity in semiconductors has been a significant factor in attracting American and Japanese inward investment.

Within the chosen advanced countries, fabrication and assembly plants tend to be located in peripheral regions where relatively cheap and pliable labour and regional development grants are available, while 'front-end' activities like marketing, design centres and development are located in or near metropolitan regions with high numbers of users. Unfortunately, many proponents of the *New International Division of Labour* thesis have been led to overlook these investments in advanced industrial countries by their concern with the runaway plants in the Third World.

The Far East

We have already noted that the role of the industry in the Far East is beginning to shift from the position enshrined in the *New International Division of Labour* thesis. The changes are due to a combination of market forces (i.e. unintended consequences of actions of individual capitals), corporate strategies and state intervention. There is a growing division between four of the newly industrializing countries – Hong Kong, Singapore, South Korea and

Taiwan – and other countries such as Malaysia, the Philippines, Thailand and Indonesia. Wages in the former countries have risen more than in the latter (United Nations Center on Transnational Corporations, 1983; Scott, 1986a). The first four countries named also have more skilled technical labour, partly as a result of improved state technical education and partly through imitation and learning-by-doing among indigenous firms, including subcontractors. The net effect of these factors is to produce an outward movement of low-skilled work on low value-added products to the cheapest labour countries and an upgrading of domestic activities and increased capital intensity in the newly industrializing countries.

The final stage of semiconductor production, testing, is now increasingly carried out in the Far East, particularly in the newly industrializing countries, having formerly been carried out in the advanced industrial countries on account of its capital and skill requirements. This change means that the finished product can be sent direct to market without being 'brought home' – a change which saves on inventory and transit costs. Further, locally owned subcontracting firms have sprung up across the region, concentrating on assembly but now including some fabrication plants in the newly industrializing countries (Scott, 1986a). Both these and foreign-owned branch plants are beginning to integrate upstream and downstream, e.g. into producing semiconductor carriers and assembling printed circuit boards respectively. Indeed, Hong Kong and Singapore, South Korea and Taiwan are now said to have the beginnings of self-generating industrial complexes in electronics, including equipment manufacturing and even centres for designing components for local industrial customers (Henderson, 1985).

South Korea's development of semiconductor production goes beyond a subcontracting role and stands out as perhaps the most dramatic attempt by a country to upgrade its position in the international division of labour. State repression of labour organization and militancy has been severe, though this on its own does not explain the rapid development of industrial capital in South Korea. What has also to be taken into account is its early advantages of a relatively developed industrial base and educated labour force together with a state strategy of encouraging not only inward investors but indigenous suppliers. From its beginnings in simple component production it has shifted away from classic late-product-cycle mature products to leading-edge components. After a six-year programme of R&D, a product targetting policy and massive state aid Samsung, the leading firm, produced 256k memories before any European firms and claimed to have undercut Japan in 64k chips (Butler, 1985). In addition

to its eleven assembly plants, South Korea now has four wafer fabrication plants, two of them jointly owned, two of them domestically
owned, and their prowess in semiconductors has been recognized
in advantageous technology transfer agreements with firms such as
Intel. Another divergence from the New International Division of
Labour, product cycle and export processing zone stereotypes lies
in the fact that the domestic market for chips has grown rapidly.
Here the presence of four large vertically integrated companies (two
of which feature in the *Fortune* list of the world's 50 largest
industrial firms) has encouraged synergies between semiconductors
and other products, e.g. helping its consumer electronics industry shift
from products like monochrome televisions to video recorders and
from thence into more design-intensive office automation products.

Such developments illustrate the more general point that although
there is a clear spatial division of labour in the industry it should
not be thought of as fixed but evolving as industries and local
economies change. Moreover, they challenge the thesis favoured by
dependency theorists that independent industrial development cannot
grow in the global periphery out of dependent industrialization.

Conclusions: redrawing the lines of competition

While the literature on the international division of labour has
emphasized changes in intra-company divisions of labour in the
semiconductor industry, in recent years shifts in the international
division of labour in terms of inter-company competition and the
shift in competitive advantage of different national industries have
been more striking.

The US companies still dominate the sector in the more design-
intensive advanced chips but are now clearly threatened by the
Japanese in at least one strategically important family of chips.
So, while Intel has a 70 per cent share of the Japanese market for
16 bit microprocessors, Japanese dynamic random access memory
(DRAM) chips now dominate the American market, though Japanese firms may yet make an impact in 32 bit microprocessors.
Europe now imports 85 per cent of its chips and over half those
produced within Europe are made by American firms. Meanwhile,
in the Third World, the newly industrializing countries are moving
up-market and a few more cheap labour countries with adequate
infrastructures and political stablility have gained production.

As is usual in technologically dynamic industries, there have
been many instances of inter-firm agreements as a means of gaining

market access or allowing integration upstream or downstream. Technological advantage and dependence are reflected respectively in licensing agreements and joint ventures such as those of European firms with US producers. But other more specific developments have prompted an extraordinary rash of collaborative agreements of varying types and degrees of formality.[14] 'Second sourcing', in which a product introduced by firm A is also manufactured by other firms, helps to spread capital costs, increase capacity utilization and meet demand peaks while reducing the need for the other firms to innovate across so many products. For example, Intel has used no less than eight second sources for its highly successful 16 bit chip.

A further kind of collaboration has involved defensive alliances of national producers set up to fight foreign competition (Organization for Economic Cooperation and Development, 1985). Rival US merchant producers have joined together through the Semiconductor Industry Association to lobby the Federal government to act to defend them, demanding better access to the Japanese market through direct investment and exports and a normalization of competition, including the outlawing of targetting as 'unfair trading'. In an attempt to provide some short-term relief the US government threatened to impose selective tariffs on Japanese chip imports because of allegations that Japanese chips were being 'dumped' at prices that did not reflect true cost. Using trade sanctions as a lever, the US government finally forced Japan to sign a bilateral chip pact. In return for the USA waiving these sanctions, Japan agreed to open its own market to overseas semiconductor producers and undertook not to dump memory chips in the US market or in other third markets. Although the pact has not been very effectively policed, its net effect may not be entirely positive for the wider US economy because, by raising the floor price of memory chips, large US users are threatening to move offshore, where anti-dumping policing is far less stringent.[15]

Well aware that short-term relief is no substitute for a strategic response, the US firms have pinned their hopes on a new semiconductor manufacturing technology project – Sematech. This proposal, estimated to cost around $1b and designed to be jointly funded by government and industry, brings together both captive and merchant producers, their materials suppliers and leading US chip users such as Hewlett-Packard, AT&T, IBM and DEC. As such, it therefore represents a degree of collaboration unprecedented in American industrial history. Unlike earlier attempts to forge cooperative ventures at the R&D end, the novelty of the Sematech proposal is that it involves a jointly operated manufacturing facility for commercial production of a new generation of DRAMs, representing a recognition that R&D

support is no longer sufficient. In short, manufacturing matters! Yet Sematech is not merely significant as an instance of collaboration, for it has increasingly taken on a protectionist bent, excluding not just foreign firms but foreign nationals working for US firms as well! (Kehoe, 1984). It has also won the support of one of the major protectionist forces in the USA, the DoD. The DoD has widened its security net around technologies and products which it considers might have a military application in Warsaw Pact countries. However, this form of protectionism raises serious issues because the USA is trying to limit the freedom of action of European firms to seek alternative sources of supply. Protectionism in the name of national security has clearly become a double-edged sword.

The American industry began with heavy state support. It quickly progressed to a period of world leadership in the process of internationalization, during which it legitimized itself by appealing to a static conception of comparative advantage and to free markets as a realm of natural justice and order. When such arguments failed to stem the rise of a new state-backed power which refused to accept the static view of comparitive advantage, it switched sides. Examples of such opportunism are of course rife in the history of capitalism, but this particular instance is notable in the light of the celebration of the alleged role of free market forces in the development of the industry in the ideology of high tech.[16]

Alliances between European firms have generally been unsuccessful as firms have considered link-ups with more technically advanced US and Japanese firms more attractive. While this may be changed by recent cooperative ventures within Europe (the ESPRIT programme, the Philips–Siemens link-up and the recent European Silicon Structures venture (involving three firms in making ASICs)), the pattern of outside connections is likely to persist given the backward and fragmented state of the European industry. These inter-company links represent responses to the extraordinary economics of the sector, to the complex and changing global pattern of protectionism and comparative advantage and, more generally, to the non-correspondence of capital and nation states. Not least, they are responses to the shake-out of firms in a still youthful industry and, in view of this, while it may seem ironic that the sector producing the heartland technology of the microelectronics revolution should be in such difficulties, it is hardly surprising.

4

Consumer Electronics

It is in the consumer electronics sector that Japanese firms have become most dominant, accounting for over half the world market; indeed the familiar products have become synonymous with the rise of Japanese industry. Prior to the entry of the Japanese, technological development in consumer electronics was relatively staid and trade and foreign direct investment were limited; since that time the sector has seen rapid internationalization and innovations such as the video recorder and the compact disc. Now, with the imminent development of 'home information systems', which are likely to be centred on an advanced version of the television, the consumer electronics sector is beginning to be incorporated into the information technology (IT) industry.

The main products are televisions, videocassette recorders (VCRs) and audio equipment. In advanced country markets, colour televisions have until recently accounted for the lion's share of the market (up to 75 per cent in Britain), with VCRs being the major new product. As a mass production industry consumer electronics implies a different set of skills from that of other sectors such as computers. Most of the production work is assembly – usually inserting components in printed circuit boards – and most of the production workers are classified as semi-skilled. As the products are standardized rather than customized, R&D activity is distinct from marketing and little or no servicing is needed – the opposite of the situation in the computer systems sector.

Production processes and products have changed dramatically in the last 15 years. Firstly, solid-state technology has replaced valve technology and the miniaturization produced by microelectronics has meant that functions formerly performed by several discrete components, which had to be individually soldered onto a printed circuit board, are now performed by chips. Consequently, the average number of components in a colour television has fallen from about 1,400 in 1968 to about 400 in 1985, with obvious reductions in the number of operators needed to assemble each set. Secondly, automation, particularly in the shape of automatic insertion machines and more recently pick-and-place robots, both first introduced by the Japanese, has further reduced the number of operator jobs. Thorn-EMI, the British firm, has found that automatic insertion reduces the number of assembly workers by a factor of seven.[1] Together, these two developments account for a massive reduction in operator semi-skilled jobs (mostly performed by women) during a time of output increase. A third change in production organization, again led by the Japanese, has been in component quality and reliability – a theme already noted in semiconductors. Western firms conventionally set very undemanding standards for acceptable quality levels for products from suppliers, e.g. a maximum of 1–1.5 defects per 100. With such permissive standards it is clear that a television set containing 1,000 components is likely to have 10–15 things wrong with it when it comes off the line. Such poor quality involves a threefold wastage of labour and materials: in producing the rejects, in identifying them and in rectifying them. By reducing the numbers of components per set and insisting on much higher quality standards – measured in terms of defects *per million* – the Japanese producers not only eliminated this waste in production but improved the reliability and hence the consumer appeal of the final product. As we shall see later, this improvement in quality has as much to do with the social organization of production as with technical changes.

Companies like Hitachi, which make both components and consumer goods, can benefit from synergies in design, by coordinating design of components and final consumer products so as to facilitate low-cost assembly (e.g. maximizing the use of automatic assembly) and to improve product performance – a possibility not open to firms such as Thorn-EMI which lack components production. The latter types of firm also become technologically passive, only able to react to changes in components technology when they reach the market and when the vertically integrated firms have already embodied them in their consumer products. However, as so often happens, many firms do not take advantage of such conditions even where they exist. Large vertically integrated firms also exist outside Japan (e.g. Philips) but the Japanese seem to have been most successful in developing the forms of internal social organization necessary to realize this potential for synergy. The Japanese companies' position is also more complicated in that, while they straddle several sectors, the organization of some production processes – televisions, for example – is highly *disintegrated*, with component production reaching down through several layers of subcontractors. But, while this is vertical disintegration in a formal sense, the *de facto* links between firms are sufficiently strong to allow the major firms to dominate the subcontract firms and gain not only the benefits of vertical integration but the advantages of greater flexibility, coordination of innovation, lower risk (to the major firms) and access to cheaper labour in the small subcontract firms. Ikeda (1979) shows how localized hierarchically organized networks of hundreds of television component subcontractors cluster around the leading firms (Sayer, 1986).

The changing global structure of the consumer electronics industry

The development of the industry can again best be explained by looking at the main core countries before proceeding to discuss internationalization. In the case of televisions, the main consumer product, the national focus is particularly appropriate because each country tends to have adopted different transmission standards. So, for example, most European countries use variants of the phased alternating line (PAL) system which differs from that used outside. These differences have inhibited exports and foreign direct investment, particularly to Europe where until recently its protectionist effects were reinforced by a licensing system for PAL standards.

In the USA, the television market was nearly saturated by 1960 (88 per cent of households possessed a television – the 1948 figure was 1 per cent). The maturing of the industry was reflected in the rapid shake-out of firms from 150 to 27 by 1960 so that, soon after, major firms like Zenith RCA and General Electric dominated the market.[2] The huge domestic market provided ample demand and this, coupled with technical differences in transmission systems and protectionism in Europe and Japan, meant that exports and foreign direct investment were limited, the former to about 1 per cent of output. However, firms like RCA did license their technology (e.g. semiconductors and tubes) to foreign firms. While this generated income it also allowed companies like Matsushita to enter the market.

As Rosenbloom and Abernathy note, the American firms treated the television as a mature product, as they had the radio, and assumed that the only avenues for change were in cosmetic alterations in styling and cost reduction (Rosenbloom and Abernathy,

1982). When competition intensified during the sixties and seventies colour television boom, with the rise of Japanese imports, the American firms responded mainly either by moving their manufacturing to low-cost labour countries like Mexico, Taiwan or Hong Kong – with consequent reductions in US workforces – or by selling 'badge-engineered' foreign products.

The approach of the Japanese firms was quite different and the result was, in Rosenbloom and Abernathy's term, a 'dematuring' of the industry.[3] From the beginning of the sixties, the government targetted consumer electronics as a key growth sector and gave it R&D grants and infant industry protection, e.g. keeping out Zenith. Despite this protection, competition between Japanese firms was keen. Until 1969, about 60 per cent of revenue came from the domestic market which although smaller than the US market was growing more rapidly. Instead of treating televisions and the like as mature products, the Japanese firms pursued both process and product innovations. So, while British consumer electronic firms spent 1 per cent of their revenue on R&D, their Japanese rivals spent 12 per cent, and that of a sales figure 11 times larger (Electronic Consumer Goods Sector Working Party, 1983); Matsushita alone spends five times the total UK budget for R&D in consumer electronics. The story is typical of the last 25 years of Japanese industry: a strong reaction against an early reputation for shoddy products leading to an obsession with quality in both process and product, reducing waste and inefficiency and improving the reliability and attractiveness of the product. Again, close involvement of management and process engineers in rationalizing production and close relationships with suppliers resulted in productivity levels twice those of the US firms and a very high utilization of capital.

These advantages, coupled with a more pro-active approach to marketing, were turned to devastating effect in penetrating foreign markets, particularly in the USA. The story of the first breakthrough is particularly noteworthy. After a major consumer survey, the American firm General Electric concluded in 1960 that there was no market for small portable black and white televisions. Three weeks later, Sony launched a highly innovative marketing campaign on small televisions and by 1965 Japanese imports totalled 1 million or one-eighth of sales of monochrome televisions. Meanwhile, American firms concentrated on colour and sold Japanese monochrome sets under their own name (Rosenbloom and Abernathy, 1982). The subsequent invasion of Japanese colour televisions with superior specifications and reliability further weakened the indigenous sector with a loss of about 100,000 jobs and led to a second bout of industrial concentration, with the number of American colour television producers falling from 16 in 1966 to 3 in the 1980s. Some of those which went out of business were taken over by their Japanese rivals: Sanyo took over Warwick, Matsushita took over Motorola's consumer activities, and Sony, Mitsubishi, Toshiba, Sharp and Hitachi all built factories in the US sunbelt. At the same time the shake-out in the industry lowered employment in the US industry from 130,000 in 1966 to 74,000 in 1978.

In Europe, the PAL licenses afforded greater protection from external competition, especially in large televisions. While there was some foreign direct investment within Europe, especially by Philips, most firms remained restricted to their small national markets, thereby forfeiting economies of scale; most European plants produced less than 150,000 sets per year, when Matsushita was making as many as a million in some of its Japanese plants. But it was not until the early seventies that Japanese imports exposed the European firms' limitations. Although in some cases slightly more technically active than American firms (as evidenced by Teletext and Prestel in the UK), European televisions were no match for the Japanese. In the mid-seventies they were hit both by imports and by the sudden end of the colour television boom, while in the eighties the expiry of the PAL licenses removed much of the remaining protection (Arnold, 1984). Now, new forms of protection, such as 'orderly marketing agreements' and 'voluntary export restraints', have taken their place, but as usual they may actually help Japanese firms by increasing the scarcity and therefore the prices of their goods. Protectionism has also induced the Japanese to invest in television manufacturing within Europe, with six plants in the UK, one in West Germany and two in Spain, though these new plants tend to import most of their components from the Far East. An enabling factor in this inward investment has been the reduction of manual assembly in television production (Cable and Clarke, 1981). While the Japanese invasion caused a major shake-out of European firms, so many have entered that the overcapacity which followed the colour television boom has persisted into the eighties.

As we shall see in Part III, the Japanese success has induced some imitation of their practices in automation and quality among European firms. However, it has not been enough to compensate for their more expensive and less flexible labour and so they remain a shadow of their former stature. The new pattern of competitive advantage is summed up by Philips, Europe's largest producer of consumer electronics: having previously transferred technology to Japan, it has now begun to import technology via its purchase of Marantz, its partial ownership of Matsushita and its technology agreements with Sony and other Japanese firms. In the Third World, Philips cut employment in countries such as India and in Latin America, while expanding it in the adopted homes of the runaways in south-east Asia and the Far East. This was coupled with reductions in its European workforce from 200,000 to 165,000 in 1980, closing down 20 assembly plants and three television tube factories, leaving it with eight television plants and one tube plant by 1985. By these means, it has rationalized its European production system so that plants are no longer geared to single national markets. Yet, while changes such as these have been emphasized in the New International Division of Labour literature, there was another aspect to Philips's strategy which was equally typical of European multinationals, namely large scale direct investment and acquisitions in North America totalling 15,000 jobs (Dicken, 1986, p. 351).

Against this we can compare Britain's largest consumer electronics firm – Thorn-EMI – which failed to adapt its more traditional style of international division of labour; its world-wide branches were mainly rental operations or plants involved in assembling kits exported from Britain. These were not intended to compete with its British plants and the company never took advantage of the runaway strategy. Unlike Philips and other leading companies, British firms saw the runaway option as a threat rather than as a potential weapon against Japan (Cable and Clarke, 1981, pp. 51, 101). This reminds us that the international division of labour, in the sense of an *aggregate, social* division (see chapter 2), is constituted not only by the forms of organization used in the most advanced firms (on which the *New International Division of Labour* and radical literature tend to focus exclusively), but by those of lesser ones to.

Now, as developed-country markets for colour televisions saturate, the front-line of competition has shifted to a new range of products in particular, VCRs and compact discs (CDs). While the video tape recorder was invented in the USA (by Ampex-Quad), and developed by Ampex and Philips, it was the Japanese who managed to turn an expensive specialized industrial product into a mass-produced consumer product. At first the Western firms ignored consumer markets. In Japan, MITI targetted it as having consumer market potential and gave a grant to Sony to develop the technology. Later, after the development of the first videocassette by Philips, eight Japanese firms worked on consumer versions. Ampex, RCA and some Japanese firms then tried out a consumer version but all failed. However, while the Western firms withdrew, the Japanese persevered and, in the case of Sony, after no less than 15 years of development work, it was eventually successful with its fourth attempt – the Betamax. The contrast with Western firms working on short-term horizons could hardly be more dramatic. Once again, this ability to hold back market forces and to invest long in advance of a return echoes the situation we encountered in semiconductors in which Japanese firms were able, through their particularly favourable relationships with the state and financial capital, to keep up R&D and investment during recessions – another example of the importance of differences in the institutional forms of capital.

As regards international competition, the situation in VCRs differs crucially from that of televisions, in that the West, with the exception of Philips, was scarcely involved in innovation and never achieved a lead in production. As with televisions, however, the assault on foreign markets came from huge highly automated factories in Japan, one of which is reputed to be large enough to satisfy the European market on its own. Now, 90 per cent of the world market for VCRs is controlled by the Japanese. While the Japanese – with the exception of Sony's Betamax – agreed on a single technical format (VHS), Philips tried to go it alone with its V2000. Although this was a highly sophisticated product technically, it has failed because of inadequate marketing. Having earlier attacked Thomson-Brandt of France for licensing JVC technology, Philips has now resorted to selling Matsushita's rival VHS model.[4] Japanese superiority is reflected in a number of other licensing and marketing agreements with Western firms, e.g. Zenith and RCA with Sony and JVC. While such licensing at least gives European firms an involvement in the market, it usually carries strings limiting exports. On the other hand, market domination or power is rarely a wholly one-sided affair and JVC's success in Europe with VHS owed much to European firms' cooperation in marketing. Combinations of competition and collaboration such as these neatly illustrate the dilemmas facing firms arising from shifting patterns of technological and market leadership and in terms of the make-or-buy decision and choice of technical standards.

European protectionism has again forced the Japanese to set up VCR plants – 16 of them – in Europe.[5] At present these are small and largely limited to 'screwdriver' assembly operations, the more critical precision-engineered work being retained in Japan. As such the plants are some way off meeting the 45 per cent local content level required to claim European origination and escape import controls. Nevertheless, the 14 per cent tariff offered little protection against imports from established Japanese firms and low-cost Korean producers.

However, at least in the CD market, Philips's position appears to be stronger, having pioneered the technology jointly with Sony. Their format has this time become the standard, and has been licensed to other Japanese firms. Nevertheless the latter have been more successful in European markets – so much so, in fact, that Philips persuaded the EEC to double the tariff on imported CDs for three years (Crisp, 1984). Now, however, all producers of CDs are threatened by the digital audio tape (DAT), a wholly Japanese innovation which matches the CD in quality but is cheaper and more versatile. The development costs of the CD have scarcely been amortized so there are problems of self-inflicted injury to their CD commitments by prematurely marketing the DAT (Rapoport, 1986). While profit margins for CDs have in any case been severely squeezed by competition, the situation illustrates a classic dilemma for capitalist firms: whether to continue to attempt to reap profits from an existing investment and to risk rival firms taking over the market with a new product or whether to write off the old and take a risk on the new.

The newly industrializing countries and a range of other Third World countries have played an important, though changing, role in the industry. The conditions for runaway strategies of shifting production to these countries first became favourable during the sixties and early seventies before the industry dematured; many American, Japanese and a few European firms took this option, moving products such as radios, pocket calculators and monochrome televisions off-shore. As mature technologies their production was largely routinized needing few skills and little management and development input, and they offered no threat to leading-edge technologies and products in the advanced industrial countries. However, since that time the scope for these classic product life cycle strategies has narrowed. In any case, the trends are complicated by the fact that the industry itself has changed in a way which (a) reduces the labour content, thereby lessening the advantages of developing-country locations, and (b), by shortening product life cycles, has narrowed the time period in which Third World subsidiaries can develop competitive production. Both these trends have slightly reduced the comparative advantage of the developing countries vis-à-vis Europe and North America. On top of this, the less developed countries lack both advanced home markets and the economic and political power to set technical standards, forcing them into alliances with probable competitors.

As we saw in the semiconductor sector, while taking some advantage of cheap labour locations, the Japanese firms also sought long-run competitive advantages in automation and product innovation. Nevertheless two countervailing factors are leading the Japanese to retain and even increase capacity in such countries. First, the rise in the value of the yen (up 35 per cent against the dollar in the mid-eighties) – itself a consequence of the success of the Japanese economy – has driven up the costs of exports, so that, at the time of writing, the wages of Japanese electronics workers were seven times those of South Korean workers. (This explains the uncharacteristic behaviour of some Japanese firms in resorting to badge-engineering Korean products!) Shifting relative wage rates between countries have again influenced the pattern of inward investment within the Third World. One recent example of this has been the diversion of some Japanese runaway television plants to Mexico, where the currency has substantially weakened against the dollar. Second, offshore production enabled Japanese firms to avoid import barriers placed against goods from Japan. Yet, as we saw in chapter 3, the use of cheap-labour countries need not rule out the use of automation; for example, Japanese firms have introduced automatic insertion machines into their Far Eastern offshore plants, albeit not the most advanced design. Nor is cheap labour the only feature of developing countries of interest to multinationals, for some have rapidly growing markets and Japanese firms have been more successful in penetrating these than have Western firms.

Some of the newly industrializing countries are moving up-market beyond providing a simple assembly function and can now produce in the early, rather than the late, part of product life cycles (Turner, 1982). The leading example of this has been South Korea which, like Japan in the sixties has targetted consumer electronics for export-oriented production. With government support, VCR production was building up to 1.8m units for 1986 at prices well below Japanese levels (de Jonquieres, 1985a). Sales in Europe and the USA have been carved up between Samsung and Gold Star. To succeed, such strategies depend not only on cheap and super-exploited labour but on a managerial lead in developing a virtuous circle of volume production and high R&D activity (Turner, 1982, p. 65). One measure of their success is the fact that over 85 per cent of television components are indigenously produced – a higher figure than many Western producers achieve (Cable and Clarke, 1981, p. 37). Another is the presence of several Korean consumer electronics plants in the USA and Europe, representing a counter-movement to traditional flows of foreign direct investment in the world economy.

As regards the immediate future, firms such as Matsushita and Sony are looking beyond consumer electronics to the expansion of IT for new markets: firstly in the terminal hardware market, particularly tapes, discs and visual display units (in which the newly industrializing countries, especially Taiwan, have been prominent) and which can represent 80 per cent of the value of an IT system, secondly in home computers, and thirdly in the development of the home information system combining audio, video, computing and telecommunications facilities, centred on an 'intelligent television' with its own microprocessor (National Economic Development Committee, 1980; Arnold, 1984; M. Smith, 1985).

Conclusions

There is no other sector of industry in which Japanese firms have been so successful in establishing themselves as world leaders, and in which the resulting changes in the structure of uneven development have been so dramatic, as in consumer electronics. This is to be expected because the sector lends itself particularly well to the exploitation of the Japanese comparative advantage in high-volume high-precision manufacturing. At the same time, the rise of the Japanese has by no means been simply a function of free market forces; indeed in many ways it is a striking example of their subordination to long-term targetting practices and their success in winning semi-political battles over standards. Similarly, particularly in Europe, their penetration of foreign markets has been shaped heavily by protectionism, obliging the firms to become 'reluctant multinationals'.

The Japanese experience has further challenged existing ideas in both academic and business circles: they have demonstrated particularly clearly that comparative advantage can be deliberately changed, that product life cycles can demature so that established and technologically staid sectors can become technologically dynamic and even propulsive, and that internationalization of production may not be essential for domination of world markets unless protectionism makes it so. However, there are limits to the extent to which the dematuring of the industry can go and many of the largest markets are now only growing slowly compared with other electronics markets. Success is never unassailable in capitalist competition and now Japanese firms are threatened by the rising value of the yen and the growth of low-cost producers in the newly industrializing countries; hence their continued success may yet depend upon their agility in diversifying beyond consumer electronics. In this respect they are considerably advantaged not only by the scope for long-term planning but by the high degree of collaboration between major firms and their subordinate suppliers regarding innovation and strategy.

5

Computer Systems

Larger in value terms than the semiconductor and consumer sectors, the computer systems industry is the most complex of the electronics sectors. It has a number of interesting features: the rapidity of technical change in the wake of the chip; the associated turbulence of economic growth; its convergence with telecommunications in information technology; the peculiarities of software and information products; and IBM'S unparalleled degree of monopoly. As before, we shall begin by outlining some key technical and economic characteristics of the industry before attempting to synthesize the global structure of competition and uneven development.

For the first 25 years of its life – up to the mid-seventies – the industry was dominated by hardware in the shape of large and highly expensive mainframe computers. In the last ten years the picture has been changed comprehensively, largely through the development and adoption of the chip, which dramatically lowered the cost of computing. New types of smaller computer – minis and micros – became feasible and have recently begun to match the more expensive mainframes in power. While the market for mainframes has continued to grow, it is minis, and especially micros, which have grown fastest. This in turn has induced expansion in the demand for computer peripherals (e.g. disc drives and printers) and for the software need to run the machines. Computer services also constitute a significant part of the industry selling not pieces of hardware and software but the *use* of these to customers. Developments in telecommunications and smaller computers have made distributed data processing possible, in which computers can communicate and share out tasks, whether through a local area network or other means such as satellites. This coupling of telecommunications and computers

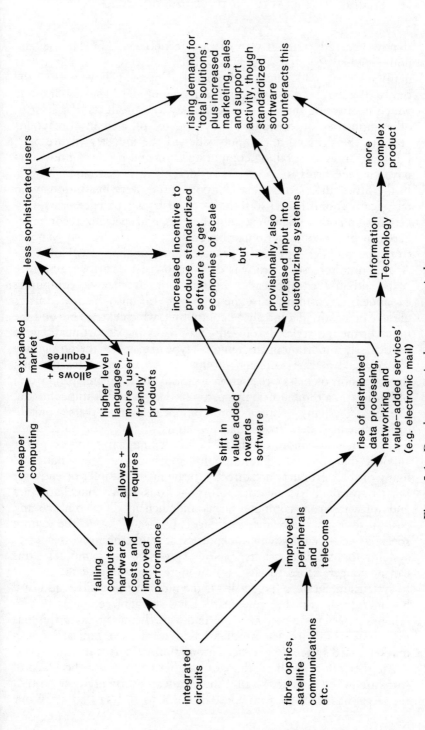

Figure 5.1 Developments in the computer industry

is now generally termed information technology (IT). Here again, both products (e.g. the automated office) and services (e.g. electronic mail) are being sold. These and other developments have been at the centre of a major transformation of the computer business, and one which is affecting the process technology and products of a host of other industries. Such changes have triggered off a chain of reactions on both the demand and supply sides of the industry (figure 5.1). On the market side, falling computing costs and product innovation have enabled firms to target a considerably wider range of buyers. In the fifties and sixties most computer users were institutions with technically sophisticated staff and specialists (e.g. defence firms, universities and other scientific establishments, and specialized computer departments or computer bureaux). As such they needed little support from the supplier and could provide much of their own software. At this time too, software was overshadowed by hardware both in costs and in the suppliers' priorities. But, as cheaper computing expanded the customer base,the proportion of unsophisticated users grew. Micros and decentralized computing, which extend beyond the limits of the old-style centralized data-processing departments, were particularly important in changing the type of user. On the supply side the enlarged markets affected production organization by allowing the extension of mass production methods for a greater proportion of hardware, although heavily customized systems are still common. Computer systems became more complicated and diverse and, in order to win orders from the new kinds of user, computer firms had to put much more emphasis on marketing, sales and customer support than hitherto. Traditionally, the main suppliers had sold many of their products not directly to the final customer or end-user but to so-called 'system integrators' – firms which buy hardware and software from computer firms and then modify, combine and customize it in ways which suit particular users (e.g. in the case of specialist technical systems integrators, it might be process control systems for chemical plants).[1] Now, however, the computer firms can no longer afford to yield up an increasing share of the market to systems integrators (especially as the number of possible 'add-ons' has increased) and therefore many have reorganized themselves to sell more systems directly to the end-user. In many cases this has meant switching from a product-based form of organization to a market-based system (e.g. office automation, factory automation).

Another effect of these changes has been to elevate the place of software in the industry so that the hardware–software cost ratio of the sector has reversed from 90:10 or 80:20 to 10:90 or 20:80. While the price–performance ratio of hardware has increased by a factor of perhaps a million, software programmers' productivity has risen only two or three times in the last 25 years (Ernst, 1985). Not surprisingly, a $5 chip can cost between $20,000 and $40,000 to programme. (Later we shall question the basis of this comparison, but the effect on cost structures is beyond doubt.) The result is that, whereas software sales used to be secondary to hardware sales, the hardware is now increasingly sold 'on the back of' software which it can run.

The resulting pressures in this area have in turn produced divergent responses. On the one hand, standardization of software is an attractive option: selling such packaged software allows the very high costs of software development to be spread over many sales, with negligible production costs, since this only involves copying onto a disc or tape. On the other hand, given their lack of technical sophistication, many users still need help in customizing hardware and software to their own needs and will choose the supplier who offers most in this respect. Similarly with micros, as numbers sold rise and as prices fall below £5,000, say, suppliers can no longer afford to provide customized software or to address potential customers on an individual basis in marketing. They are therefore increasingly turning to dealers and retailers to sell them, thereby producing divergent trends towards both standardization and customization

While the chip and other technologies have enabled these developments, the demand side has not merely been passive for, as always, new products create further needs. Today, most customers are not interested in technical specifications of hardware but want a solution to a specific set of practical problems that they have. Inevitably, employing a single firm or consultant which can provide a total solution is preferable to working it out oneself and approaching a range of suppliers. All the trends in figure 5.1 converge in this rising demand for 'total solutions'.

These trends have compounded the increasing demands already put upon computer firms by the convergence of computing with telecommunications in IT. Consequently, no firm can expect to produce and service every requirement. Firms offering the widest range will have an advantage but they will inevitably have to buy in some elements of each solution. Both these and other firms have also had to develop capabilities in providing solutions for particular 'vertical markets' such as systems for supermarkets. These trends have both opened up numerous niches – often filled by new firms – and obliged established firms to collaborate to cope with the new developments.

The types of work and occupations found in the computer industry differ markedly from the stereotypes of manufacturing which we mentioned in chapter 2. Unless this is appreciated, little sense can be made of either competitive strategies or the structure of uneven spatial development at global or subnational levels. As in consumer electronics, automation and ICs have reduced the labour involved in hardware production and assembly. But in any case the numbers of such workers have always been relatively small. Even before the shift towards greater customer orientation, much of the industry's labour force was involved not in the direct production of hardware but in R&D, software and in interacting with potential and existing customers; indeed, follow-up sales of add-ons, software enhancement, consultancy, user training courses and contract maintenance may generate more revenue than the original sale. Hence it is common for well over half the employees of computer firms to be in marketing, sales, service and support (House of Lords Select Committee on Overseas Trade, 1985). Moreover, such employment profiles are as likely to be found in foreign subsidiaries as in domestic operations – contrary to the familiar radical stereotypes of branch plants; for example, no less than 37 per cent of IBM France's 21,000 strong workforce are of executive status (Bakis, 1980).

The need for interaction also derives from the nature of IT products and services, even where technically sophisticated customers are concerned. In comparison with most commodities, the use values of IT products are unusually numerous, open-ended and loosely defined. Whereas it is relatively clear what the possible uses of, say, a press or drill are, the range of uses of a computer system or piece of software such as a spreadsheet package is enormous. This is only partly due to the 'immature' nature of the product: it is also inevitable because information is not information unless it is capable of very large numbers of permutations. Even the manufacturers of such products are unlikely to appreciate the full range. However, sophisticated users such as telecommunications firms may come to know more about their possibilities and limitations through learning on the job and hence become a source of innovations, some of which may be of commercial interest to the computer supplier. In such cases buyer–seller interaction is likely to be intense.[2]

We referred earlier to the productivity disparity between hardware and software. One obvious problem with such comparisons is a confusion of labour productivity in producing hardware with the performance of the resulting hardware. But there are other problems too, and to resolve them we must consider what goes on when someone produces software. Normally, once a hardware product has been designed for the market it can be produced in whatever volume is required, using a standard set of techniques, and the manufacturing process itself – or 'production' – is likely to involve a considerable number of the labour force of the firm, usually far more than are involved in design. Software is different. Even though a large proportion of *sales* may consist of standard packages, which are simply recorded from master copies, the bulk of the *work* involved in the industry consists of developing new products or adapting and maintaining (i.e. correcting) old ones. Certainly, in many cases, the work may be quite routine and boring, combining a fairly standardized set of procedures in a slightly novel way. But it involves a greater element of design and problem-solving than does most production – problems which are likely to become more difficult over time – and sometimes it may deserve to be classed as R&D. On these grounds alone, then, it is unreasonable to compare productivity in software production with labour and capital productivity in hardware.

Three sets of influences have structured competition and the social organization of the industry: first, the market and product developments already described; second, the remarkable power of the industry leader, IBM; and third, the effect of technical standards and complementarities. The degree of monopoly or size of market share possessed by IBM in the computer industry is without parallel. It is the market leader (and often the profit leader) in every capitalist country except Japan. It is eight to ten times the size of its nearest rivals, the 'BUNCH' (Burroughs, Univac, NCR, Control Data and Honeywell, also all US-owned), 16 times the size of the largest European producer, Siemens, and over 40 times the size of ICL, the largest British-owned company. IBM's research budget alone is greater than the total annual revenue of Bull, the leading French computer firm. In view of its dominance it is scarcely an exaggeration to say that IBM is not so much a competitor to other firms as the environment in which they operate.

The issue of technical standards concerns the fact that different producers tend to use incompatible standards for hardware and software although it is not uncommon for the same company's products to be incompatible!). In particular, each tends to have its own proprietary operating system (the software which actually runs the computer's basic functions). This means that users become 'locked in' to a single supplier because their often extensive inventories of applications software (word-processing, payroll, accounting, data base packages etc.) will not run on another firms' equipment (Mackenzie, 1985). A second area of contention of growing urgency concerns the standards governing the linking-up or networking of computers and other telecommunications and office equipment. Many other computer firms and foreign governments are anxious to establish their open systems interconnection (OSI) as the standard in opposition to IBM's systems network architecture (SNA) which, if adopted in Europe, will further deepen the European firms' technological dependence. While initially SNA looked the stronger, determined opposition has forced IBM to pay homage to OSI. Such problems are likely to be exacerbated by the convergence of sectors in IT, as this requires the harmonization of standards in computers, telecommunications and office equipment.

The conjunction of this kind of problem with IBM's dominance is evident in the difficult choice facing other producers regarding whether they should adopt IBM technical standards so as to give 'plug compatibility' and to imitate IBM machines, or go it alone with independent standards and specifications. The independent strategy eases the pressure from IBM but until recently was less attractive to users. The plug compatible strategy gives better access to existing IBM users and gives non-IBM users access to IBM's vast software range. It is also made easier by the fact that today key components can be bought cheaply, and this has opened up niches for firms making cheap imitations of IBM's (or Apple's) personal computers (PCs) in low-wage countries. On the other hand, the plug compatibles not only help to sell IBM software but are vulnerable to price competition, since IBM can spread its costs over a higher volume. They are also threatened by the introduction of new IBM products, though some Far Eastern firms have lessened this problem by reducing the imitation time lag to six months.

However, the diversification of the industry and the effect of the microprocessor in lowering barriers to entry has created new openings. The expansion of the industry in the USA has seen waves of new innovative firms, each associated with new families of products, e.g. Digital Equipment Corporation (DEC) and Data General with minicomputers and, most famously, Apple with micros. Meanwhile, the BUNCH, and other old firms, have found it increasingly difficult to survive. Squeezed by IBM but lacking strength in minis and micros, they are merging (e.g. Burroughs and Univac) and diversifying to reduce their dependence on mainframes.

The new firms generally produced more technically advanced products than IBM but lacked the latter's enormous customer base and sales and support force. Many purchasers wanted to use minis and micros as additions to existing IBM set-ups. At first, problems of incompatibility reduced the attractiveness of the new firms' products. Yet IBM was slow to react to such changes and suffered a fall in market share in the mid-seventies (from 60 per cent in 1967 to 40 per cent in 1980). It then came back with a vengeance. Building on its scale advantages, it undertook a programme of factory automation costing over $10b, re-equipping 42 plants in 14 countries, making big strides in penetrating some rapidly growing specialist markets such as computer-aided design and manufacturing equipment and launching a micro, the PC, which became the market leader in only two years. Significantly, the latter was technically rather pedestrian but thanks to IBM's reputation and back-up, coupled with price-cutting and aggressive marketing, it outsold more advanced machines such as the Apple Macintosh. Despite the rapidity of technological change in the industry's products, ability to produce and market efficiently and to provide long-term customer support counted for more. In two respects, the development trends of the industry favoured IBM: the breadth of the company's product range had become a greater advantage as large systems of linked computers became more popular, and its traditional strength in marketing meant that it did not need to change as much in this direction as more technology-driven companies. As a result, IBM increased its share in world mainframe and micro markets to 76 per cent and 45 per cent respectively by 1984.[3]

However, as usual, technologies and markets do not always turn out to be tractable, even for large firms, and, like many companies, IBM has recently run into difficulties with its technology for networking computers and with its strategy of locking in customers – so much so that it has actually had to cut its workforce for the first time. As the growth of computer hardware and software products outstrips the ranges of even the biggest suppliers, the disadvantages to users of being locked into a single supplier's products increase. Consequently firms like DEC which offer not only compatibility between their own products but compatibility with other firms' products are now coming back against IBM which offers neither. Such are the twists and turns of the industry's technological and market development.

The diversification of the market and the increased complexity of computer systems with the trend towards 'selling solutions' have obliged firms to collaborate in order to be able to offer products outside their existing expertise. As traditional boundaries between sectors and product areas are sapped, a remarkable rash of inter-firm agreements has resulted. Even IBM has had to do this to keep up with new developments; for example in semiconductors it has an equity stake in Intel and it has bought into Rolm, an innovative US telecommunications firm. In addition to creating enormous technical and organizational difficulties, convergence also opens up new sources of competition as companies which traditionally restricted their business to telecommunications try to invade computer markets. And herein lies what is likely to be the most serious challenge to IBM's dominance, in the shape of the US telecommunications giant AT&T, as we shall explain in chapter 6.

We have already noted the tendency of the mass media to see the electronics and computer industry as immune to recession and we have seen how mistaken this is in the case of semiconductors. Although not as violently cyclical as the semiconductor industry, the growth of the world computer industry has slowed down in recent years for many firms. One of the peculiarities of the industry is its combination of markets which have matured (mainframes) alongside others which are apparently in the infant industry stage (e.g. micros – at least until recently – and perhaps software). As infant industries mature there is normally a shake-out of all but the strongest companies, but in computing this has been particularly severe and rapid. In the case of micros, hundreds of firms have failed and the industry looks to have matured about four years after its birth. At the end of 1984 there were still 158 firms offering business micros but the majority of these are expected to drop out soon as just ten of them already have 60 per cent of the market.

The crucial factor here has been the ability of IBM to move over into the micro market, speeding up the shake-out and threatening even companies as vigorous as Apple. As the industry matures, technological change tends to reduce the number of alternative standards and designs, thereby simply cutting out many early entrants. The 'upstarts' that do develop advanced products have difficulty finding a place in a market already overcrowded with scores of different products. Lacking their own distribution channels, they have to rely on being accepted by distributors and retailers, who are only capable of selling a restricted range of products to customers. (In the case of home computers, there is literally a restriction of shelf space in the shops.) Consequently only a small proportion of home computers and software packages ever reach a market of any size and those that do are immediately faced with the problem of gaining a mass production facility in a short period of time. Then, if they can cope with this success, they have next to come up quickly with another competitive product if they are not to be left behind. Most fail.

The other factor exacerbating the shake-out is the recession or slow-down in computer markets since 1983. While this is certainly related to the wider economic recession, as firms cut back on their capital spending, there is generally agreed to be another factor. This is the widespread experience among customers of difficulty in learning how to make effective use of their existing purchases and delays in achieving the hoped-for savings. As computer/information systems have become bigger and more complex, the upheavals and learning curve problems involved in their introduction into institutions have increased, and this too has raised consumer scepticism and resistance. In some cases there have been actual falls in demand and many of the most lauded new generation firms have cut their workforces: Data General by 1300; Wang (previously increasing its revenue by over 50 per cent per year) by 1,600; Apple by 1,200 in 1984–5; while Hewlett-Packard has introduced a 5 per cent pay cut. Again, this is further evidence against the belief that the industry is recession-proof.

However, we do not wish to give the impression that there is no longer any room for new ventures. New niches will surely continue to open up for new firms to fill almost as quickly as old ones are eliminated or swallowed up by big league firms. The creation of a major new software market with the introduction of IBM's PC was one such example. On the other hand, even the largest firms can run into difficulties in adapting to IT products, as indeed IBM has. Overall, it seems that the following industrial structure is emerging:

1 IBM plus AT&T, and one or two Japanese firms supplying basic mass markets in hardware and some software;
2 a larger tier of systems integrators supplying vertical markets and creating custom systems for niche markets (many of the medium-sized European firms are moving in this direction);
3 a host of small specialized suppliers providing the sectors described in point 2 with individual pieces of hardware and software tailored to specific markets (adapted from *Business Week* (1984)).

Computers: uneven spatial development

The USA has dominated the computer industry from the start. Like semiconductors its early development was heavily dependent on military contracts at a time when commercial markets were uncertain. The famous ENIAC, built in 1949 and one of the first computers, was funded by the US Army Ballistics Laboratory. Again, the size of the American market and the military/space programme connection have been decisive; the USA is just over half the world market while the Department of Defense (DoD) is the world's largest user of software. A further advantage of the USA is the synergy already noted between computers and semiconductors.

Unlike the situation in semiconductors and consumer electronics, European countries lagged from the start in computers, having been hampered by their small and relatively backward national markets. In Britain, for example, the Ministry of Defence, unlike the US

DoD, chose not to fund the development of computers but bought products already available on the market. ICL, the British 'national champion' in computers, never benefitted from R&D funds under the defence budget, as did Honeywell, IBM and Burroughs (Hills, 1985). The big American firms established themselves in the European markets soon after the war and now IBM alone accounts for over half of European output.

The larger firms generally tried to be 'mini-IBMs' at first. While their products were frequently technically more sophisticated than IBM's, they could rarely match its marketing power, and what market share they did gain against IBM and the BUNCH owed much to government protection, particularly in the form of preferential procurement. In Britain this was for many years sufficient to keep ICL, the biggest indigenously owned firm, ahead of IBM in market share, though not in terms of profit. (On joining the EEC, the UK had to abandon this policy, though it had in any case been limited by political opposition from foreign multinationals resident in Britain, such as Honeywell.) The rise of minicomputers and micros and the European weakness in ICs stretched these firms still further, as leading US firms poured in during the seventies – firms like DEC, Hewlett-Packard and Commodore. In response to this invasion most European countries have tried to follow a 'national champion' policy, but this has proved wholly inadequate because the national markets are too small to allow competitive performance and the policy simultaneously inhibits foreign direct investment and exports within Europe.

As a result of their weak position with respect to the technological and market trajectories of the industry, the European firms are having to buy in more technology and products by making deals with over-seas companies and are having to become more like systems integrators. However, while in the latter respect they have the advantage of local knowledge of markets, in entering alliances they have much less leverage than American companies. For example, ICL markets Fujitsu mainframes as well as its own, and buys 80 per cent of its chips from Fujitsu. This is not to say that some medium-sized and small firms have not prospered (e.g. Olivetti of Italy and Nixdorf of West Germany) but their role has been largely a secondary one in the industry, in groups 2 and 3 of our classification. Meanwhile, the much-vaunted European producers of home micros are rapidly being eliminated by a shake-out and a slow-down in demand. Their employment creation effects have in any case been modest (e.g. about 600 workers to assemble 100,000 Sinclair micros per month) and most of the parts are bought in, many from the USA and Japan. Finally, European firms are weak in the lucrative area of computer peripheral equipment.

In Japan, computers were targetted in the seventies when MITI grouped six leading firms into three pairs in order to challenge IBM. As usual this was quite unlike a national champion policy for, while the firms collaborated on pre-production research, competition in the market among the six was nevertheless fierce and by 1979 Fujitsu had overtaken IBM in market share. The Japanese firms have had most success in selling mainframes (mostly plug compatible), portable micros and particularly peripherals such as printers and VDUs, which currently account for about 80 per cent of the sector's exports. However, as regards complete systems, they have generally not matched their success in semiconductors and consumer electronics. The main reasons for this are, first, that except for mass-produced precision goods like printers the industry gives less scope for the exploitation of Japanese manufacturing strengths, second, that curiously, given Japanese prowess in manufacturing, the market for office automation is very backward and, in addition, networking was banned until recently, and third, that the industry is backward in software, particularly in standard packages and in customer support, this being a major deterrent to potential buyers of its hardware. However, what software they have produced has a reputation for being highly reliable, thanks in part to the use of quality circles. The complexity of the Japanese language has had a double-edged effect: while it has made the development of software difficult, providing it in Kanji characters reduces foreign penetration of the domestic market. And, while progress towards changing their comparative advantages and disadvantages has been slow in this case, the expensive fifth generation programme for the development of 'intelligent' computers has demonstrated their commitment to Western rivals (Arnold and Guy, 1986).

Given the nature of the sector – especially its cost structure and need for contact with customers – it is not surprising that 94 per cent of the non-communist world production by value is located in just six advanced countries (USA, Japan, West Germany, France, Britain and Italy). As yet, only a few Third World countries such as Taiwan and Singapore have been involved in the industry as assembly bases for higher volume 'commodity' products, such as monitors, which need no interaction between producer and buyer. Assembly of smaller products produced in high volume has generally gone to low-wage, often assisted, areas in the European and North American periphery (e.g. Ireland, Puerto Rico) and more recently to Far Eastern countries such as Taiwan for visual display units (VDUs). One American firm, Tandon, which makes floppy disc drives, is now shifting all its manufacturing to India and the Far East (*New York Times*, 22 April 1984). As with semiconductors, such instances tend to get most exposure because they are most likely to involve cuts in production capacity in the advanced industrial countries. However, according to McEwan, the rule that computer firms invest overseas mainly to penetrate local markets rather than for cheap overseas export-oriented production facilities still applies to the majority of foreign direct investment in the Third World. Import controls, local content agreements and proximity to customers – particularly the state – are still important in such contexts:

> Within Latin America, for example, US investment in manufacturing has become increasingly concentrated in Brazil and Mexico, the nations with large, rapidly growing markets, not the nations with the cheapest labour. (McEwan, 1985).

Foreign direct investment by European and Japanese firms reflects their greater distance from leading-edge technology and markets in that they are obliged to play a bootstrapping role, for there is still a one-year time lag in the spread of state of the art American products in Europe and Japan. However, with the exception of some cheap plug-compatible computers – by definition not state of the art – and sales of Japanese computer peripherals, foreign producers have made little impression on the American market.

Such is the size of IBM that we can explain a significant part of the global distribution and structure of the industry by outlining its internal organization of production internationally, though IBM's famed secrecy about its operations is a handicap in this task. Globally, its domination of virtually every non-communist national market has posed the danger of nationalization or exclusion by foreign governments. This has strongly affected IBM's locational strategy. In response, it has developed an elaborate system of complementation, i.e. a network of manufacturing plants each specializing in just a few parts of computer systems and linked by large cross-border flows of materials and products. No one computer system is made entirely within one country (which reduces the threat of nationalization) and imports and exports to and from each country are kept roughly in balance so that the company cannot be charged with upsetting the balance of payments. This inevitably reinforces the tendency of foreign direct investment to be concentrated near markets. For example, having previously been responsible for a wide range of products for the UK and European markets, the Greenock plant in Scotland now assembles input output devices and terminals. Various peripheral equipment, including disc storage systems, is assembled and tested in Havant in Hampshire. Certain mainframes and personal computers are assembled in Boca Raton, Florida. Printed circuit board assembly and testing take place in Austin, Texas, and Bordeaux, typewriters are made in Lexington, Kentucky, telecommunications equipment manufacture takes place in Montpelier, large minicomputers and disc drives are assembled in Virnecate near Milan, discs are made in Mainz in West Germany and point-of-sale systems are made in Santa Palomba in Italy (IBM Company Reports). And so the list goes on – more recently supplemented by an increase in subcontracting and buying in parts, such as VDUs, from Taiwan. The result of such extensive foreign direct investment is that, in many countries, IBM employs more than the largest indigenous producer and therefore can plausibly claim domestic producer status and thereby increase its political influence. For example, in France, with four manufacturing plants, two R&D centres and several other operations, plus a workforce of 21,000, IBM has at last been accepted as a domestic producer after years of discrimination by the government.[4]

IBM's huge R&D budget, which exceeds that of the US space programme, is also distributed across a large number of specialized but linked operations. Research is by no means entirely centralized in the USA; for example, there are R&D activities at Hursley in the UK (disc storage systems), Zurich (process technology) and Nice, France, (satellite communications and fibre optics). According to Bakis (1980), each of these overseas operations is paralleled by an equivalent specialist laboratory in the USA with which it cooperates. Again, the reasons for this decentralization are partly political but there are other reasons too, and ones which other major producers share. One is the need to relate R&D to local customers' evolving needs and to monitor developments which might have more general application. Another is the global shortage of skilled labour. Labour markets are especially tight in major high-tech areas such as California, and in the USA as a whole 15 per cent of higher-level electronic engineers are of European or Third World origin (Ernst, 1985). But American firms are not relying wholly on 'brain drains' particularly as know-how is closely related to local markets and technology. For these reasons, major firms can scarcely afford not to have foreign research laboratories.

The importance of marketing and customer support in the industry and especially in IBM is reflected in its numerous service operations scattered throughout all its major markets, the more routine activities forming a 'cloning' structure (Massey, 1984) similar to a central place network, the more specialized formed on a complementation basis.[5] While this kind of activity has tended to be neglected in industrial location studies, perhaps on account of the small size of most units, they are nevertheless the most common type of foreign set-up and are an essential element in its competitiveness. Where these point-of-sale branches customize systems for users their contribution to the final value of the product may be as significant as that of manufacturing and should not necessarily be thought of as an inferior kind of investment (National Economic Development Committee, 1983b).

Enormous amounts of information are generated both by the need for continual interaction with customers and by the complex product range and corporate spatial division of labour. In IBM, this in itself requires a network of internal information processing centres for coordinating activities around the world; for example, Cosham, near Southampton, Uithoorn in the Netherlands and Boulder, Colorado, are just three elements of a vast network of data-processing centres linked by the most advanced telecommunications. As Bakis (1980) shows, this network enables the company to circulate relevant information from any part of the system to virtually every operation in the world within 24 hours, thereby reducing response times and greatly increasing the utilization and turnover of capital.

Conclusions

Computer systems are undergoing particularly dramatic change with the rise of IT; indeed the whole identity of the sector is changing. By comparison with the semiconductor and consumer electronics sectors it has a much stronger market orientation and a more decentralized distribution of front-end activities. Similarly, with the exception of certain peripheral equipment (notably VDUs), it has made much less use of the runaway option. Like semiconductors, this is a sector in which American firms have retained the lead in product technology while losing it in volume production of hardware to the Japanese. The effect of this in terms of imports of peripherals and semiconductors was a trade deficit in IT with Japan of $17.6b in 1985 (Kehoe, 1986). Meanwhile Europe is in deficit in its trade with both Japan and the USA and is likely to remain so while its firms are restricted to more specialist markets and to gaining what leverage they can from outside suppliers looking for alliances.

6

Telecommunications

The boundaries of the telecommunications industry have been under siege for some time as computers, office equipment and telecommunications converge into the so-called information technology (IT) sector. Nevertheless we believe that telecommunications deserves to be addressed in its own right, for the convergence is still in its infancy and the sector remains distinctive, particularly in terms of its massive dependence on protected national markets.

For most of this century telecommunications services have been operated and controlled by a public monopoly in most countries, the one major exception being the USA where regulated private companies run the service. The prevalence of the public monopoly has been attributed to a number of factors: national security considerations, the 'natural monopoly' of the telephone network, the public service character of the telephone service and the need to maintain the technical integrity of the network. As a result one of the key features of the telecommunications market, at least until recently, is that it is dominated on the demand side by a public monopsony and on the supply side by a national 'club' of private manufacturers. In stark contrast with semiconductors, consumer goods and computers, telecommunications markets have been severely partitioned along national lines, a result of chauvinistic public procurement policies and a thicket of nationally specific technical standards. This set-up afforded each national club of suppliers almost total protection against both foreign and domestic competition. With the exceptions of defence and agriculture, no other industry has received such privileged treatment from the state. Accustomed for so long to a quiet life and a rich diet of guaranteed profits from public contracts, the traditional suppliers are now having to face up to three new challenges:

1 Converging technologies enable firms from bordering sectors to enter telecommunications markets, though this is proving far more difficult than many expected.
2 Product markets are becoming more and more internationalized, partly because burgeoning development costs often need more than one national market if they are to be amortized.
3 Institutional change in the shape of deregulation is slowly prising open protected national markets; and the countries that have pioneered deregulation (the USA, the UK and Japan) are demanding reciprocal access to other markets.

Telecommunications networks consist of three basic types of equipment and each has witnessed rapid technological change over the past decade or so. As regards switching equipment, traditionally the major segment because it accounts for some 50 per cent of the value-added in the industry, the key change has been the transition from electromechanical analogue systems to electronic digital systems. Modern digital systems are essentially special-purpose computers. As a result, the traditional suppliers have had to acquire new expertise, especially in software and integrated circuits (ICs); in fact software alone is estimated to account for some 60 per cent of all R&D costs (Organization for Economic Cooperation and Development, 1983). It is in the sphere of public switching systems that soaring development costs are felt most acutely, since it now costs between $500,000 and $1b to develop a new generation of digital switch (Commission of the European Communities, 1984b). Hardly surprising, then, that collaboration and takeovers are now rife in this segment.

In the transmission sphere the most important trend has been the broadening range of transmission media. Conventional (copper) cable is giving way to combinations of optical fibre cable, microwave and satellite systems. Finally, as regards terminal equipment (which is normally located on the user's premises, e.g. telephones and private automatic branch exchanges (PABXs), there has been a massive upsurge in demand for equipment that can handle data, text and images as well as basic telephony. This upsurge has been driven in the main by the precocious growth of the private business market and this has reduced the service providers' share of equipment purchases (Organization for Economic Cooperation and Development, 1983). Taken together, these innovations signal the emergence of a new telecommunications infrastructure as existing services can be enhanced and new generations of service are on the horizon.

The supply of telecommunications equipment is highly concentrated, the top ten firms accounting for some 80 per cent of world sales. Until fairly recently, it was possible to distinguish between two groups of firms: a majority group which was geared to protected national markets and a smaller group which had a global orientation forced upon them because of the limited size of their domestic markets; Ericsson of Sweden is the outstanding example in the latter group. But this distinction is fast becoming obsolete since most if not all of these firms are now obliged to address international markets. Over the last few years the spate of mergers, joint ventures, licensing agreements and foreign direct investment bears witness to this new era. Indeed, even the largest firms cannot hope to supply and service the entire spectrum of product markets on their own given the nuances of national markets.

Although telecommunications is invariably cast as a growth sector the impact of microelectronics in the equipment part is inducing jobless growth, displacing large numbers of semi-skilled workers but boosting the employment of scientists, electronics engineers, software specialists and systems analysts (Organization for Economic Cooperation and Development, 1983). The main employment growth in telecommunications will tend to be found in new generations of telecommunications services.

The USA

Nowhere has technological and institutional change combined with greater effect, at home and abroad, than in the USA. Although the USA is the largest telecommunications market in the world, its biggest producer – AT&T – did not begin to address itself to overseas markets until the momentous divestiture battle was resolved. Divestiture, which was agreed in 1982 and took effect in 1984, reduced AT&T to around a quarter of its former size, since the Justice Department forced AT&T to relinquish its 22 Bell operating companies. But in exchange AT&T was allowed to manufacture for and compete in markets (like computers and semiconductors) from which it had been legally excluded. Significantly, on the very same day that the divestiture settlement was announced (8 January 1982) a 13-year anti-trust suit against IBM was finally dropped – a signal, no doubt, that anti-trust restrictions were being relaxed so that US firms could face external competition, particularly from the Far East.

Prior to divestiture, AT&T bestrode the US telecommunications sector like a colossus. With assets of $150b and annual revenues of $69b AT&T was the largest corporation in the world, providing end to end services which included R&D (Bell Laboratories), manufacturing (Western Electric), local services (the Bell operating companies) and long-distance services (Long Lines division). However, as a regulated utility, AT&T was forbidden from entering other markets (such as computers and semiconductors) although Bell Laboratories was free to develop any technology for in-house purposes. More importantly, AT&T was obliged to license all patents controlled by Bell Laboratories to any applicant at a 'reasonable' royalty. Given the fact that the Laboratories had registered 20,000 patents by 1983, or one per day since 1925, it fulfilled the function of a national electronics laboratory, geared to applied R&D. Such has been the significance of Bell Laboratories in inventing and diffusing information technology that some of its managers have claimed that 'without Bell Labs there would be no Silicon Valley' (Office of Technology Assessment, 1986).

AT&T's extraordinary domination began to be assailed in the late 1960s. Aided and abetted by a new wave of technological innovation, AT&T was subjected to anti-monopoly pressure from business users and aspiring suppliers. Together, these forces secured a series of deregulatory measures which allowed new firms to compete with AT&T in both equipment and long-distance service sectors (Brock, 1981; Irwin, 1984). AT&T found itself ensnared in an unenviable conundrum: it was forbidden to enter bordering markets such as data processing, yet unregulated firms from these bordering sectors were able to enter its fiefdom. From the late 1960s onwards AT&T began losing market share as a result of its higher prices and the fact that it was relatively slow in introducing new electronics-based products. However, its performance was inextricably tied to its regulated status: it had always designed and manufactured equipment on the assumption that this equipment would have a long, reliable and protected life and new products were subject to a lengthy regulatory review. Unencumbered by these constraints, its competitors were able to introduce products whenever they chose. One of the clearest examples of the new situation was in the field of digital PABXs, where AT&T's share fell from 100 per cent in 1976 to 24 per cent in 1985 as a result of better and cheaper products from Northern Telecom, Rolm and Mitel (Office of Technology Assessment, 1986). Further inroads into AT&T's domination were made by imports from cheap labour countries of commodity items such as telephone handsets in which labour costs are a critical factor. In these lower-end market segments many US firms are moving offshore, the most dramatic move being that of AT&T's domestic telephone production from Louisiana (itself a cheap labour area within the USA) to Singapore in 1985 to exploit labour-cost differentials.

The US telecommunications market remains the largest, the most liberalized and the most technologically advanced in the world.[1] Not surprisingly it is now the most targetted national market in the world. One reason for this is the advanced nature of semiconductor suppliers in the USA. But the attractions lie on the demand side too, in the form of synergies with advanced innovative users. In view of this need for supplier–user interaction, it is virtually impossible for imports to penetrate the US market for the more sophisticated products.However, what the US firms find particularly irksome is the fact that they have not been able to translate their technological muscle into commercial success in overseas telecommunications markets because these are politically protected. Acting on behalf of US firms the Federal government is desperately trying to overcome what it perceives to be a glaring anomaly, namely that trade surpluses correlate not with economic efficiency but with political protection. The US government feels that its relatively 'open-door' policy has helped to propel the USA from a surplus to a growing deficit in telecommunications trade, especially with Japan and the Far East which together account for around 75 per cent of all US telecommunications imports. (What is less publicized, however, is the fact that this deficit partly reflects the 'unpatriotic' behaviour of US plants exporting to the USA from the Far East!) Paradoxically, the growing US trade deficit in telecommunications is largely the result of two state policies: unilateral liberalization and the divestiture of AT&T. Freed from AT&T's in-house procurement policy, the Bell Operating Companies–which are aggregated into seven regional holding companies, each comparable in scale to British Telecom–became keen to diversify their purchases away from AT&T, thus creating hitherto unknown opportunities for foreign suppliers, especially in basic equipment. In an attempt to offset these unintended consequences of deregulation and divestiture, the Federal government is further relaxing anti-trust restrictions within the USA. At the same time it is trying to prise open overseas telecommunications markets through direct political pressure in order to secure what the US authorities refer to as a 'level playing field' in the international telecommunications market (Schiller, 1982, 1983).

At the corporate level the likes of AT&T and IBM need no political prompting to pursue global strategies, and such is their size that their strategies condense some of the wider trends now under way in the emerging IT sector. The backgrounds and corporate traditions of these two firms are radically different. IBM, an accomplished multinational, is pre-eminently a marketing-driven organization. AT&T, a novice in world markets, was a technology-driven organization until its divestiture. Despite their colossal size and the aura of invincibility that surrounds them, both have been forced to acknowledge that they cannot diversify into each and every IT segment without a series of alliances and acquisitions at home and abroad. In the telecommunications field, now a high priority with IBM, the acquisition of PABX manufacturer Rolm was one of the most tangible signs of the convergence of telecommunications and data processing, since the digital PABX is widely seen as the central nervous system of the electronic office, the hub to which all voice and non-voice equipment will be attached. However, in its further moves into the telecommunications field (by takeover of Satellite Business Systems (SBS)) IBM has largely failed because it underestimated the problems of marketing advanced voice and data communications services through failing to prepare the market. Its subsequent decision to merge SBS with MCI, which served the more mundane long-distance telephone service market, represented an admission of defeat, reminding us that even the largest actors cannot move in and out of sectors as easily as is sometimes thought unless they have the requisite skills.

For AT&T there are two main challenges: how best to diversify into new product markets and which strategies to adopt to penetrate overseas markets given its lack of expertise outside the USA. Thus far its response has been broadly the same in both cases, namely a combination of alliances, foreign direct investment and accelerated product and process innovation, the latter being one way of reducing its high-cost structure. AT&T's keen appetite for international alliances bears witness to its belief that size and technological prowess are not of themselves sufficient to penetrate overseas markets. Three of these alliances deserve to be mentioned here.

1 Its joint venture with Philips in public telecommunications equipment is essentially a marriage between AT&T's technology and Philips's world-wide marketing network, particularly its political network for accessing regulated telecommunications markets in Western Europe. The geographical remit of this joint venture speaks volumes as to where the power lies in this alliance; the designated territory includes Western Europe, Africa, Latin America and the Middle East. In other words AT&T has reserved the most lucrative markets for itself, namely North America, Japan and the Pacific Basin.

2 Its acquisition of a 25 per cent equity stake in Olivetti, one of Europe's foremost data-processing and office equipment suppliers, is more ambitious than the one with Philips. Here, in addition to distributing each others' products in their respective home markets,

they will cooperate on joint development of new products. For AT&T the main attraction of this link was that it provided access to Olivetti's marketing network in Europe, especially to Olivetti's intimate knowledge of the European office automation market, where the norms and standards are unlike those in the USA.

3 It has entered a joint venture with Mitsui to supply data communications services in the newly liberalized Japanese telecommunications market.

As well as forging alliances to build up its underdeveloped overseas marketing network, AT&T has also begun to internationalize its production and R&D facilities. Overseas production sites include Singapore, as we have seen, and a joint venture with CTNE, the Spanish telecommunications group, to build a microchip plant in Spain. More significant was the decision to open the first overseas subsidiary of Bell Laboratories in Japan, an acknowledgement of Japan's emerging leadership in applying advanced technology in its IT industries. Overall, however, AT&T has been chastened by its entry into the competitive arena: since 1983 it has lost around $1b on its computer activities and, in a desperate drive to cut costs, it has shed over 80,000 jobs since 1984. The mixed experiences of both IBM and AT&T illustrate that capital's capacity for 'hyper-mobility' is sometimes exaggerated, not just in the geographical sense but also in the sectoral sense.

We have already seen how, far from stunting innovations, the regulations placed upon AT&T made its Bell Laboratories a prolific source of inventions for the USA's electronics industry as a whole. Now that the company has been deregulated many fear that a 'privatized' AT&T will no longer allow Bell Laboratories to function as a quasi-public R&D laboratory, liberally diffusing its innovations to other US firms. Furthermore, it is claimed that this problem, coupled with foreign penetration by imports and inward investment, will produce negative multiplier effects because the US telecommunications industry has traditionally been a creative first user of advanced microelectronic components (Borrus et al., 1984; Business Week, 1985b). Ironically, while such fears proliferate in the USA, Western Europe is transfixed by the thought that US deregulation will see further US penetration of its own IT sectors.

Western Europe

Telecommunications occupies a distinctive position in the European Community (EC) countries because it is the only IT sector in which the EC has a net trade surplus, although, in recent years, this aggregate picture conceals mounting deficits with both the USA and Japan. Yet this strength is deceptive since it is partly a function of the protected status of the European telecommunications industry. European equipment compares unfavourably with that of US firms, prices for broadly similar equipment being 60–100 per cent higher for switching systems and 40 per cent higher for transmission equipment (Organization for Economic Cooperation and Development, 1983). Equally disturbing, the EC is particularly vulnerable in some of the core technologies, such as ICs and data processing, which account for a growing proportion of the value-added in telecommunications products. Consequently, the EC now imports around 83 per cent of the microelectronic components used in telecommunications equipment (Commission of the European Communities, 1984a).

Forming 20 per cent of the world market (as against 35 per cent and 11 per cent for the USA and Japan respectively), the EC is fragmented into no less than 12 different network traditions, 12 different industrial systems and 12 different systems of regulatory practice (Ungerer, 1986). None of these markets constitutes more than 6 per cent of the world market. This fragmentation is a major problem as spiralling development costs need to be spread across ever larger numbers of users if they are to be amortized. For example, there are five indigenous European digital switching systems available, each of which costs between $700m and $1b to develop and each of which needs a market of some $14b if this is to be recovered – far larger than any national market in Europe (McKinsey & Co., 1983).

European firms have been sustained in the past by the special relationship which they enjoyed with their national network operators, e.g. British Telecom in the UK, the DGT in France and the Bundespost in West Germany. Although the operator share of the total telecommunications equipment market is declining, relative to the private business user share, it remains enormous in absolute terms, accounting for at least 70 per cent of the market in each EC country. As we have seen, the chauvinistic procurement policies of these national carriers is the major reason why the EC telecommunications market remains fragmented along national lines. Now, somewhat frenetically, the Western European countries are trying to adjust to the new trends in the telecommunications sector. Belatedly, the European Commission has been trying to promote a supra-national strategy for telecommunications since 1983. The Commission argues that the limits of national markets and the associated duplication of resources are the most serious obstacles to the EC's ability to defend itself against further US and Japanese encroachment.

The 'imperatives' in the telecommunications sector may be broadly similar for all the EC countries but, thus far, what is most striking is the diverse ways in which each country is negotiating them. By and large this can be explained by two factors: (a) divergent national state strategies and (b) the uneven record of the national telecommunications clubs. Here the UK and West Germany offer the most conspicuous contrasts, even though both are ruled by 'Conservative' governments. Under the Thatcher administration the UK has pioneered a liberal telecommunications regime, so much so that the UK has the most deregulated market in Europe. In addition, British Telecom's monopoly over the network has been broken, a second carrier (Mercury) has been licensed and British Telecom has been privatized, with the result that it feels less obliged to support its traditional national suppliers. No other EC country has ventured so far down the liberal path. The justification for this path was said to be that the UK's record was so much worse than that of its EC partners, a point which carries some force even if we discount the role of ideological dogma. However, one of the main effects of the Thatcher strategy is that the UK telecommunications firms – GEC, Plessey and STC – are now the least protected in the EC. Despite its 'social market' image West Germany operates the most illiberal telecommunications regime in the EC. This situation is buttressed by a broad-based political consensus in favour of the Bundespost's monopoly and by the fact that the telecommunications club, centred around Siemens, has performed relatively well by UK standards. Overall there is a greater degree of 'industrial patriotism', from the state for the telecommunications industry in France and West Germany than there is in the UK (Morgan and Webber, 1986).

Despite their different degrees of health, all the European firms share one fundamental objective, namely to establish alliances. Unlike the offensive convergence-seeking alliances of IBM and AT&T, mergers such as those of Alcatel and Thomson in France, Italtel and Telettra in Italy, and GEC and Plessey in the UK are primarily defensive in nature, being designed mainly to share costs and reduce duplication of resources. Gaining access to the US market now absorbs much of the energy of the European firms. The most successful has been Siemens, the largest telecommunications firm in Europe and, significantly, the least interested in pan-European alliances. Siemens was the first EC firm to break into the prestigious and notoriously competitive Bell market in the USA. In addition to this it has also secured the most far-reaching alliance in the USA, merging its US telecommunications activities with those of GTE, the largest vertically integrated telephone company in the USA. Siemens's ultimate aim here seems to be a marriage between its own technology and GTE's telephone companies which, if successful, would provide Siemens with a captive equipment market in the USA. With the exception of Siemens and Ericsson, European firms have found it exceedingly difficult to penetrate the US telecommunications market. The entrenched position of North American firms such as AT&T, Northern Telecom and IBM–Rolm and their ability to use their already installed equipment base to lever further sales in the form of enhancements, add-ons and peripheral equipment still presents formidable barriers to entry. A second barrier is the heavy charges associated with the stringent US approval procedure together with the cost of adapting European systems to US technical standards which can amount to some 30 per cent of the original development cost of a digital switching system (Ungerer, 1986). For most European firms the prospect of breaking into this open market is proving to be a chimera, with the result that they are effectively left with two options, namely collaboration or withdrawal. Not surprisingly, it is the collaborative option on which the European Commission has chosen to base its supra-national strategy for telecommunications.

The Commission estimates that telecommunications, defined in terms of equipment and service segments, will become the EC's major income-generating sector by 1990. Because of this, combined with the indirect effects of the telecommunications sector on the wider economy, the Commission concludes that the EC 'must stake its all' on a Community strategy for telecommunications (Commission of the European Communities, 1984b). This involves trying (a) to construct a common market by promoting Community standards and, ultimately, a Community-wide procurement policy and (b) to sponsor collaborative R&D programmes. Elegant as this supra-national strategy may seem in principle, it faces some acute practical problems. At the corporate level it is clear that many EC firms perceive each other as 'hostile brothers', with the accent on 'hostile', while collaborative ventures between EC and US and Japanese firms outweigh wholly European alliances by a ratio of three to one (Ungerer, 1986). At the political level sharp differences are already evident as regards deregulation: while the French are the major proponents of coordinated European action, the UK government has effectively transferred control over the UK's telecommunications space to the market. On top of this, it is clear that the supra-national strategy is medium to long term in nature, while the corporate strategies of the EC firms are dictated by immediate threats from their more powerful US and Japanese rivals. Here the weight of non-EC alliances indicates where these EC firms perceive their interests to lie. However, if the EC countries are unable to reverse their faltering position in telecommunications their failure will exact a terrible cost. The telecommunications industry is a relatively more important end-user of ICs in the EC than in either the USA or Japan and, by the end of the century, it is estimated that 60 per cent of the European IT sector will be heavily conditioned by telecommunications (Carpentier, 1985). Any weakness here would therefore further erode the EC's tenuous grip in semiconductors and data processing. For this reason it seems likely that political intervention will prevent the shake-out of suppliers which would otherwise occur.

Japan

If the EC's position is not what it was, Japan has emerged as a major world force in telecommunications, a feat often obscured by its more visible successes in consumer goods and in semiconductors. Today, however, Japan boasts the second-largest national telecommunications network in the world and this, perhaps more than anything else, has been used to lever Japanese firms into positions of strength in international telecommunications markets. Much of the credit for this developmental programme must go to NTT, Japan's national network operator. Until 1985, when it was privatized, NTT functioned as a public R&D centre like Bell Laboratories and as a powerful procurement agency. Without a manufacturing capability of its own, NTT procured the bulk of its equipment from a family of four firms, NEC, Hitachi, Fujitsu and Oki. Initially, these firms utilized imported US technology, availing themselves of AT&T's liberal licensing policy. But, with the formation of NTT in 1952, collaborative programmes were set in motion to develop an indigenous technological base. Whereas British collaborative programmes tended to stifle innovation – mainly because of cosy cost-plus contracts and corporate secrecy – the Japanese managed to develop a highly effective infrastructure for inter-firm R&D collaboration. In contrast with its British counterpart, NTT enforced a far more rigorous regime of incentives and sanctions in its procurement contracts – testament, once again, to the different forms that state intervention can take (Okimoto and Hayase, 1985).

For most of the post-war period Japan's telecommunications industry grew mainly on the strength of domestic demand. However, a dynamic relationship appears to have evolved between the home market and the export strategies of the NTT 'family' firms. For example, it is argued that

NTT has helped to develop and finance pilot and mass production systems for manufacture of the products jointly researched and developed. Crucially, NTT has procured high volumes of equipment at premium prices from its family companies – not unlike the US Department of Defence – which serves both to make demand highly predictable and stable, and to subsidize price competition for those Japanese firms on export markets. Of course, all these developmental activities have been closed to foreign firms. In essence, NTT's industrial policy role has enabled favoured Japanese telecom–computer–semiconductor companies to develop and commercialize new technologies in a protected, subsidized and risk-minimized way. (Borrus et al., 1984)

Protectionism and subsidization in Japan and Europe have therefore had completely different results. The Japanese firms seem to have made greater use of the learning-by-doing experience acquired in the domestic market to propel themselves into export markets and, later, into foreign direct investment. However, this can work both ways. Significantly, it seems that, where Japanese firms have not had the domestic market experience in installing and debugging equipment, they have encountered serious problems in overseas markets.[2] They lag behind the USA, for instance, in software, digital public switching and digital PABXs but have strong positions in transmission equipment (e.g. optical fibre cable and microwave), ICs and an array of other products where the stress is on hardware quality. However, large R&D efforts are under way to overcome areas of traditional weakness and these will make Japan a credible threat in the spheres of US strength. In fact a recent US assessment concluded by saying that 'if present trends continue, the component technologies on which the information age depends will be dominated by Japan. It may then also follow that Japanese companies will ultimately dominate the design and manufacturing of the telecommunications systems that are based on these technologies' (*IEEE Spectrum*, 1986). Unlike most of their European competitors, the major Japanese telecommunications firms already excel in base technologies like microelectronics, and their size and integration has enabled them to internalize much of the technological and commercial convergence, thereby reducing the need for external acquisitions.

No mention of convergence in the Japanese context can avoid reference to NTT's ambitious integrated network system (INS) project. Over the next 15 years NTT aims to create a broadband communications infrastructure, in which hitherto separate networks – for telephony, facsimile, data and video – are integrated into a single digital network. While it has been estimated that this INS project could cost over $80b, INS-related markets are expected to total over $200b, over and above NTT's direct investments (*Eurogestion*, 1986). Clearly, INS constitutes much more that a telecommunications project. It is, rather, a communications processing system which serves as an integrating mechanism for telecommunications, computing, office automation and home electronics equipment. No other country appears to have such a well-developed vision of the leverage which telecommunications can have over other IT sectors. Even if it is not fully implemented, the INS infrastructural project seems likely to provide unparalleled opportunities for Japanese IT firms to establish strong positions at home and abroad (Arnold and Guy, 1986).

However, new imponderables have entered the Japanese telecommunications system as a result of external political pressure from the USA and internal bureaucratic pressure from MITI (Hills, 1986). In consequence, the Japanese telecommunications market is being liberalized, NTT's procurement policy is slowly being opened up to foreign tenders and, in 1985, NTT was privatized. In this new situation the traditional alliance between NTT and its suppliers is breaking down, at least by the organic standards of the past. Nevertheless, the long lead times involved in R&D projects like INS make it unlikely that the collaborative relationship between NTT and its family firms will dissolve overnight, besides which, the extent of the family firms' dependence on NTT should not be exaggerated. Even if NTT remains a critically important market for these firms, particularly in the early stages of the product cycle, their current dependence on NTT appears to be slight. As far back as 1981, NTT accounted for only 12 per cent of the total sales of NEC and Fujitsu, falling to a mere 2 per cent in the case of Hitachi (Okimoto and Hayase, 1985). Unlike UK firms, who are far more dependent on national procurement, the Japanese have used their protected domestic base as a platform for exports and overseas direct investment. Indeed, Japan's growing telecommunications trade surplus with the USA – around $1.8b in 1986 – is an index of its emerging comparative advantage, and this cannot be entirely reduced to its excessive protectionism.

Conclusions and prospects

By now, the distinctive nature of the telecommunications sector should be clear. We have tried to outline the main trends now under way in telecommunications, emphasizing three in particular, namely changing state policies and environments, collapsing sectoral boundaries and the ever-growing need to break out of national markets. In no other part of electronics is capital accumulation so heavily dependent on political mediation and internationalization so limited. Even liberalization itself requires not a simple 'hands off' decision but active state intervention, including the pressurizing of foreign governments to reciprocate in opening up their markets. In this process pro-liberal governments (like the US and the British) have teamed up with multinational telecommunications users to force the pace of liberalization in those countries (such as West Germany) which have been slow to open up their internal market. However, although trade accounts for a small share of total output and foreign direct investment is only beginning, international markets and production structures are at last emerging. Consequently, the factors behind competitive success in the other electronic sectors – technological innovation, quality-conscious manufacturing, financial resources and marketing networks – will have to be taken on board by telecommunications firms if they are to survive in this less-protected environment.

However slow, it seems likely that, as with computer systems, the form of internationalization will be dominated not by examples of the runaway industry phenomenon but by advanced country locations. Within the advanced countries, we have argued that the prospectus is far from auspicious for EC firms, with the exception of Siemens. Most EC firms are weak in the base technologies; they have yet to make much impression in the USA and very few are even attempting to penetrate the Japanese market. Given their reluctance to collaborate, it would seem that, at the corporate level, further rationalization of the EC equipment industry is inevitable. Yet, in view of the strategic significance of this sector, it is extremely unlikely that indigenous firms will be allowed to go to the wall despite all the economic 'imperatives', and despite political regimes like Thatcherism. To illustrate the political significance attached to telecommunications we can do no better than to reiterate the Commission's conclusion that the Community must 'stake its all' on this sector because it is the key to growth and control in the wider IT complex.

7

Electronics in Global Perspective

Two major themes have run through part II – the changing international division of labour and the nature of competition and strategy in relation to companies and states. Having looked at four major sectors in some detail we can now draw out the theoretical implications of these themes, enlarging upon our initial discussions of theory in chapter 2. But first we shall begin by summarizing the overall global position.

In terms of general patterns the relative strengths and different roles of the sectors in the main countries or blocs for the sectors can be discerned from trade and employment data. In the case of trade, as shown in Figures 7.1 and 7.2, the differing strengths of the Japanese and American industries in consumer electronics, as against 'informatics' (office machinery and electronic components), can readily been seen although allowance must be made for foreign direct investment and the consequent non-correspondence of countries and 'their' firms. Particularly clear are the Japanese dominance of consumer electronics, Japanese resistance to imports in both categories and the extensive use of 'rest of the world' locations, especially by American firms. (It should be remembered that the 'rest of the world' includes not only the Third World and the newly industrializing countries, but countries like Australia, Canada and the Comecon bloc.) In electronics as a whole, the US had an $8.6b trade deficit, accounted for not only by its huge deficit with Japan, but by imports from the rest of the world; in the latter case, Singapore, Taiwan and South Korea accounted for $1,312m, $2,493m and $1,517m respectively (Electronics Industrial Association, 1986). However, since many of the imports from such countries are from subsidiaries of US companies and since exports to third countries from these subsidiaries are not distinguished, the position of American multinational companies is much stronger than the trade figures indicate. Similar qualifications apply to European and Japanese firms.

Also surprisingly clear is the reflection of developments at the global level in employment change (figure 7.3). Even allowing for the fact that the categories used by each country are differently defined, certain consistent patterns come through which are plausible in the light of the foregoing account. Taking electronics as a whole the weakness of the European countries can be seen, together with the vigour of growth in South Korea. Japan's conquering of world consumer electronics markets is reflected in the fact that it is the only advanced country to have consistent employment growth in this sector, while Europe and America have suffered heavy losses.[1] With a few exceptions, such as the West German computer industry, employment in Europe in the other sectors has stagnated or declined slightly. Finally, bearing in mind the caveat about data comparability, the figures give a rough indication of the relative sizes of the main sectors in leading countries, the relative insignificance of consumer

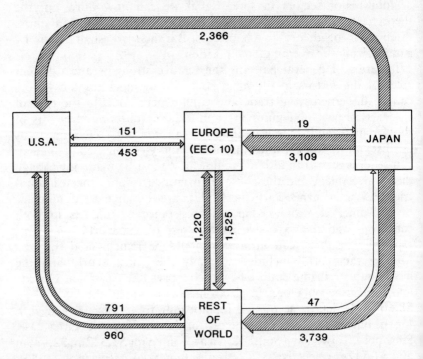

Figure 7.1 Trade flows in consumer electronics, in millions US$ – 1982 (based on data from Ludolph, 1984)

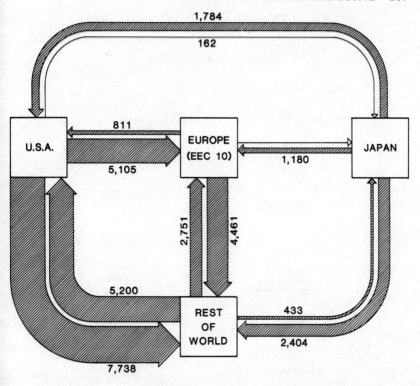

Figure 7.2 Trade flows in 'informatics' (office machinery and electronic components), in millions US$ – 1982 (based on data from Ludolph, 1984)

electronics in the US industry, its importance in Japan and the small size of the computer industry in Europe. One thing which the figures do not show but which was evident from the preceding chapters is that growth has occurred not just through expansion of markets for given products but through a rapid increase in the range of products and hence a deepening of the social division of labour. This runs counter to radical accounts which tend to focus on the expansion of markets for a fixed range of products and hence only on the expansion of technical divisions of labour.[2]

The most striking thing about global patterns of trade and foreign direct investment in electronics, as in most other industries, is the overwhelming dominance of the advanced economies as destinations – even allowing for the runaway industry and newly industrializing country phenomena. This reflects not only the technical sophistication of the product but the need, strongest in the computer and telecommunications sectors, to interact with customers. Another

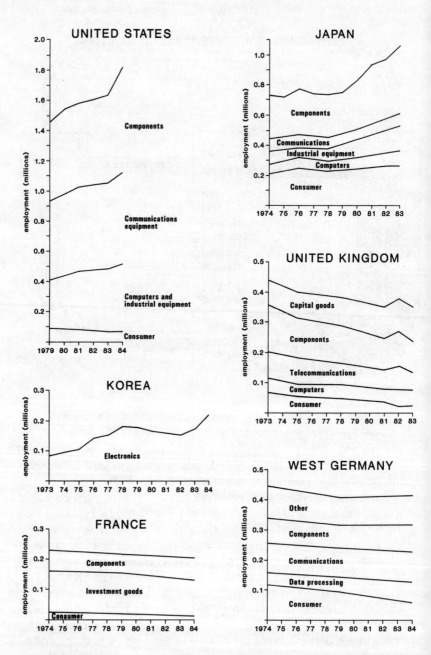

Figure 7.3 Electronics employment in selected countries (based on OECD statistical data)

consideration is the avoidance of protectionism, whether actual or threatened, and this has certainly affected sectors like consumer electronics and standard components which might otherwise have relied more heavily upon exports. However, the picture of rich country dominance is changing as the European industry continues to fall back, as Japan has conquered mass production and threatens to move into more advanced product design beyond consumer goods, and as the newly industrializing countries threaten to become 'new Japans'. The latter cases have increasingly shifted beyond the export-processing zone stereotypes, only producing late-product-cycle goods such as black and white televisions. Through a combination of imitation, learning by doing through subcontracting, upgrading of labour skills and strong state intervention, several countries are succeeding in following a bootstrapping policy.

What kind of new international division of labour?

As the radical literature emphasizes, the driving force behind all these patterns of internationalization is capital accumulation, not technological change and diffusion on its own, and major companies cannot expect to survive and to continue to accumulate capital while remaining within the confines of even large domestic markets. However, they can expand by several different means. The experience of Japanese firms shows that, under certain conditions (a highly favourable domestic environment for production and high-value, low-weight mass-produced goods), an export strategy may be superior to one based on foreign direct investment. Lack of internationalization of production therefore need not be read as an index of backwardness. Where foreign direct investment does occur, it is usually in response to contingent circumstances such as import barriers or the need for close access to customers. One trend regarding foreign direct investment which does seem to be common, however, is complementation (see chapter 5) and this has greatly increased the integration of plants into global systems of production.

From what we have seen, there are three major organizing principles behind the uneven spatial development of industry at this level:

1 the legacy of existing forms of uneven spatial development, including national and local influences upon capital accumulation, such as government–industry relations, wage rates and markets (recall that although the spatial form of product markets tends to be less coherent than labour markets, they are still spatially specific);

2 the internal organization of activities within major firms (corporate spatial divisions of labour are dictated by this in combination with the first principle);
3 the spatial form of competition within the industry, structured not by the internal division of labour of Philips or whatever, but by competition *between* companies for markets, e.g. Philips versus Hitachi. While, increasingly, the markets of such firms inter-penetrate, there is still a significant degree of spatial monopoly and, as we have seen, multinationals generally still have most power in their home market. This is due not simply to the protection of distance in terms of travel costs but to the entrenched nature of buyer–seller relationships.

Our emphasis in the explanation of these patterns and changes has been microeconomic or mesoeconomic, but macroeconomic influences have of course been present too, as traditional development theory has always stressed. One obvious example is the effect of Japan's considerable success in export markets on the value of the yen, which has made foreign direct investment more attractive to Japanese firms. The alternative response of raising domestic wages in order to expand the domestic market has met with more than the usual resistance because this would worsen the large companies' export positions by pushing up costs. But the effect of low wages and salaries on capital are of course double-edged, not only because they allow only limited purchasing power for consumer goods but because they reduce the need for automation and hence the market for many kinds of electronics product.

We have laid stress on the home bases of the multinationals (a) because these serve as launching pads for their exports and foreign direct investment, (b) because multinationals often enjoy preferential treatment in their home markets, and (c) because multinationals still have a national identity derived from the characteristics of their home economy. This is evident in every electronics sector and is largely responsible for the huge contrasts between US, Japanese and European firms.

Although this kind of analysis is influenced by radical theory regarding the mechanisms of uneven development, it is partly at odds with its assumptions about the effects of those mechanisms. One of the most important contributions to this theory has been work such as that of Hymer which attempts to relate the internal structure of multinationals to the emerging global division of labour.

... a regime of North Atlantic Multinational Corporations would tend to centralise high level decision-making occupations in a few key cities in the advanced countries, surrounded by a number of regional subcapitals, and confine the rest of the world to the lower levels of activitiy and income, i.e. to the status of towns and villages in a new Imperial system. Income, status, authority and consumption patterns would radiate out from these centres along a declining curve, and the existing pattern of inequality and dependency would be perpetuated. The pattern would be complex, just as the structure of the corporation is complex, but the basic relation between different countries would be one of the superior subordinate, head office and branch plant. (Hymer, 1972, p. 38)

While such ideas have been fruitful, we can perhaps now see their limitations more clearly.

Apart from the underestimation of the role of Japan, the more obvious shortcomings concern the challenge to the hierarchies from the newly industrializing countries, the limited extent of decentralization to peripheral countries and the persistence of advanced countries as leading production bases, the limited and uneven (though increasing) amount of foreign direct investment in some sectors and the effect of nation states as both constraints upon and actors within the international division of labour. Yet, as we have already hinted, there is a deeper problem with this kind of explanation of uneven development which is a tendency to stress the second of our structuring principles – the internal organization of activities within firms – at the expense of our third – relationships between firms, whether in the form of competition or of different patterns of specialization.

Another problem with the radical literature in this field is a tendency to underestimate the possibilities for *change* in spatial divisions of labour, whether at the international or subnational levels; according to Hymer the effect of internationalization would be that 'the existing pattern of inequalities and dependence would be perpetuated'. This view lies at the opposite extreme from the traditional non-structuralist literature, in which corporate activities are conceptualized in abstraction from the larger structures of interdependent activities of which they are part. This permits the common fallacy that many areas can simultaneously 'clone Silicon Valley', which we discussed earlier. But on the radical side there is the danger of assuming that spatial divisions of labour are fixed or that, if change does occur, it can only involve individual firms or localities changing places within that fixed structure. It was indeed clear that firms reproduce particular local labour market characteristics (e.g. low-paid labour in the Far Eastern countries) and that often the continuation of their activities in those localities is dependent on the reproduction of those conditions. But, as we have seen, some *are* changing their place (Japan, Hong Kong,

South Korea etc.), whether intentionally, as in the case of training labour where there are skill shortages, or inadvertently, as in the case of upward drift of wages in high growth areas such as Silicon Valley or Hong Kong. Such developments certainly take place within the wider structure of the international division of labour but they are also helping to change the nature of that structure. It is not like changing seats on a bus, for the nature of the seats and the transport are changing too, though without collapsing into an unstructured mass in which individuals can do anything, regardless of others.[3]

Radical accounts of global shifts in industry have not concentrated just on the division of labour but have tended to treat labour as the prime determinant of that division. Extrapolating from limited cases such as those of standard electronic components and clothing, it is assumed that markets can be accessed from anywhere. Within this framework, cheap labour and the runaway industries have dominated attention to such an extent that it becomes difficult to see why an industry which is so highly internationalized should be so heavily skewed towards the advanced countries.[4] Traditionally, before the operations of multinationals were taken into account by authors such as Hymer, it was common to underestimate the extent to which production locations could be divorced from major markets. We have now to avoid going to the equally unsatisfactory opposite extreme and take due account of market location and spatial monopolies. The failure to resolve this problem is in no small measure due to the assumption that labour is always the main source of capital's problems. As we have repeatedly seen, getting access to resistant industrial and public sector markets is often a more critical factor for firms, particularly in electronics, and such matters may have more influence on their profitability than efficiency in production. Similarly, this recalls our point from chapter 2 that not all corporate spatial divisions of labour arise primarily in order to take advantage of different kinds of labour; for example, foreign telecommunications firms have extended their spatial divisions of labour into Britain in recent years not because of the special nature of British labour but to exploit the opportunities of the liberalization of the UK market. Alongside the international division of labour there is an uneven international development of markets and, despite the common and often considerable separation of the locations of production from consumption, the former is profoundly influenced by the latter. Obvious though this may sound, the *New International Division of Labour* thesis has encouraged many to overlook it.

Certainly, cost minimization motives are important in investment decisions, but the geographical consequences depend upon the nature of the products, as was clear in the contrast between mass-produced goods such as consumer electronics and standard ICs and more specialized products such as telephone switching systems.[5] There should also have been sufficient examples in the foregoing chapters of apparently similar products with quite different cost structures to give the lie to facile expectations of simple one-sided shifts in the international division of labour.

Competition, strategy and the state

Several of the points we have made so far suggest that simple models of firms and competition common in economic theory are deficient. It is not just that price competition based on process innovation or cost lowering is supplemented by competition through product innovation, important though this is. It is the fact that competition is rendered hazardous by the long lead times of development, by the uncertainties of choice of technical standards as technological and market trajectories change and by the interdependencies of competitors' actions that makes strategy so important. In addition, the picture is complicated by the non-correspondence of nation states and capital and by a variety of forms of state intervention.

To elaborate, let us take the nature of firms first. Multinationals tend to be coalitions of interests, of varying degrees of integrity; they are not simple monoliths. They embody several different technical divisions of labour, not one, and are not simply hierarchical in organization but are matrix-like. Thus it would not be rational for them to behave in a unilinear fashion; the risks and opportunities they face are often transient and therefore most companies spread their risks rather than back a single line of attack. This in itself helps to account for the 'messiness' of aggregate geographical patterns of industrial operations.

The kind of thinking we have criticized has tended to be associated with a view which exaggerates the strategic power and internal coherence of individual multinationals in the world economy *vis-à-vis* their sectors and national governments. Carefully selected examples can of course lend support to this emphasis and in some of the radical literature the multinationals' power is almost celebrated. One feature of this view is a tendency to interpret the actions of capitals retrospectively as the largely successful execution of grand strategies.[6] Yet multinationals frequently behave opportunistically in response to unforeseen and localized circumstances. Consequently, many parts of the internal spatial divisions of labour of multinationals are the product of *ad hoc* action. Even where multinationals do attempt to put major corporate strategies into action, they frequently go wrong (e.g. the 'World Car' strategy in the automobile industry). This is not simply because the future is unknowable or because there are often unrecognized problems with top-down planning (overruling the knowledge of those at the bottom etc.) but because of the crisis tendencies endemic in capitalism, particularly crises of overproduction. However big the multinational, however great its role as a generator of market forces, it is also subject *to* them. Changing technical trajectories and market trends present capital with huge uncertainties and the behaviour of multinationals consequently often resembles groping in the dark rather than the competent execution of long-run comprehensive strategies. When deciding how to act in relation to these trajectories, companies are obliged to start from where they are rather than from where they would like to be, or from a neutral position, as economic models assume. For some this may mean technological leadership, for others technological dependence. In either case the constraints of existing fixed capital consist in more than sunk costs but in particular technological forms, such as operating systems in computing, or transmission standards in televisions and telecommunications, which lock them into existing ways of operating. Some companies will have advanced in-house suppliers or users, others will not, some will have good access to markets through distribution and dealer networks while others lack these, and similarly with respect to access to scientific labour or cheap unskilled labour. All of these characteristics are of course relative and, in many ways, interdependent too. What Philips decides to do is based on judgements about what Sony et al. might do, and vice versa. And, as Philips knows to its cost (with the V2000 video format), where choice of technical standards is concerned market success or failure can be an all-or-nothing affair. Furthermore, many markets are intensely politicized and competition for them may largely take the form of pre-production political bargaining over access and technical standards e.g. that between the EEC Commissioner and MITI.

Interdependencies of these sorts make strategy critical, but equally, in view of the inevitable uncertainties and problems of even the best-laid plans coupled with the difficulty of coordinating large organizations, one can understand how opportunism is also common, not only among weak companies fishing for short-term profits but among major companies faced by glaring unexpected opportunities. When confronted with this shifting complex of trade-offs and possibilities, even Japanese firms, with their long planning horizons, may behave opportunistically, and hence 'out of character', reminding us once again that economic forces are stronger than the largest and most powerful firm.

The interdependencies between rival corporate actions give competition a complex game-like structure which in any situation invites several quite different responses, many of them quite rational. Short-run and long-run interests may lie in different directions, as in the case of AT&T's dilemma regarding the effects of liberalization in aiding competitors. In these kinds of situations, then, one should expect radical differences in response, even in quite similar conditions. This comes out in the double-edged nature of so many of the circumstances affecting the evolution of firms. Protectionism can induce either lethargy in the protected firms or, if combined with other circumstances as happened in Japan, a rapid build-up of infant industries to a strength where they become major forces on the international scene. Similarly, large domestic markets can be an advantage in providing a launching pad for attacks on foreign markets through exports and foreign direct investment, as happend in the US semiconductor industry, or they can again induce lethargy and disinterest in overseas markets as happened with US television firms.[7] There are even cases where the smallness of the home market has prompted the development of a well-designed strategy for selling abroad, as in the case of the Swedish telecommunications firm Ericsson. In other words, the conditions relevant to the evolution of capitals in different places do not operate singly and in a consistent manner, regardless of other circumstances. Their influence depends upon how they are combined with other conditions. This, however, is nothing more than a variant of a more basic problem: people and institutions are of course capable of interpreting and responding to the same conditions differently, even when not confronted with such complex trade-offs.

A further complication to competitive processes is added by collaboration. The thickets of collaborative deals between electronics firms, often of different nationality, bear witness to the struggle for survival in an industry undergoing both rapid internationalization and technical change. Link-ups reflect the need to penetrate foreign markets, to use spare capacity in a situation of overproduction and to offset the effects of the discrepancies between firms' capabilities and market and technological trajectories as developments in electronics sap the boundaries between traditional sectors. Increased size and financial power are not the only considerations in successful alliances. Where firms need to forge alliances to cope with the problems of technological convergence, as in IT, the choice of a partner with the appropriate specific technological capacities is critical. This is often insufficiently appreciated, especially in the British context mania' where many observers share with much of British capital the illusion that the material bases of exchange value are unimportant. A corollary of this is that convergence of sectors such as computing and telecommunications is far from a smooth process of growing similarity of products and overlap of markets allowing capitals to join harmoniously and ease across into new fields. In addition to the huge problems of technical incompatibilites, there are also incompatibilities of a social and organizational kind. Telecommunications operating companies like AT&T and British Telecom come from a regulated market, with a paternalistic management style and a highly unionized workforce. The data-processing companies have never been regulated and, for the most part, their workforces are not unionized. The resulting clash has already produced many casualties in terms of redundancies, as we saw in the case of AT&T, and given that convergence is generally occurring during a time of liberalization it is all the more likely that the industrial relations of the telecommunications sector will be conquered by those of the computer sector. Convergence therefore often seems more like collision; not surprisingly, even some of the most powerful 'convergers' are now recoiling painfully form their first encounters.

There are then the relationships between capitals and states to consider. The contradictions between capitals' interests and those of national economies are well known in the case of developing countries and foreign multinationals, yet they are also present in advanced countries. In addition to the differences of interests, which are universal, the problems of non-correspondence between nations and companies have been especially acute in Europe with its small domestic markets, its many 'national champion' firms and conflicts between EC members. In different and less acute ways the dilemmas of US liberalization in telecommunications highlight the problems. Similarly, US trade deficits are swollen by imports from American offshore production yet, without these, American capital would have more problems in competing with the Japanese. Only in Japan do the contradictions between capital and national economy appear muted, thanks to limited internationalization and inward investment and the country's unique institutional arrangements between state and capital. However, even here, the problems of the rising yen have begun to disrupt this harmony in recent years.

From our analysis of the four main sectors of the electronics industry it is clear that we are dealing here with highly politicized markets, so much so that the conventional juxtaposition of 'state versus market' or 'government versus industry' miserably fails to capture the manifold ways in which the state shapes this industry and constitutes its markets – indeed, the state is itself a large market! State-sponsored R&D programmes, tariff and non-tariff barriers to entry, political intervention to secure access to foreign markets and public procurement are the most important examples of this process. This is true whatever the political regime and whatever the country. The fact that these interventions have been more successful in some countries than in others is due as much, if not more, to the environment in which they are carried out as it is to the form of intervention, a point which is too often overlooked.

In the USA and Japan there has been both strong state intervention and powerful and innovative capital and both countries have benefited from large internal markets. But the forms of intervention and of capital's own organization have been utterly different in these countries. In the USA the Department of Defense, NASA and the Buy America Act provided vital support for vigorous firms organized in a loose vertically disintegrated structure. In Japan, active state intervention established new firms where needed, encouraged and subsidized cooperative R&D and provided market closure and support for exports. Large vertically integrated firms used their tight internal organization and relationships within industrial groups to forge a formidable manufacturing capability.

Meanwhile, in Western Europe, state interventions have foundered on small markets, political fragmentation and the interpenetration of and dependence upon foreign capital, while few firms have been equipped to use state support to become more rather than less competitive. At the beginning of the book and in chapter 2 we noted the tangled webs of identities and conflicts of interest between firms and between governments and firms that can arise where national markets are highly penetrated by foreign capital; nowhere is this better illustrated than in Western Europe. Both governments and large firms are actors in the process of uneven development, to be sure, but the relationships between them are exceedingly complex.

PART III

Electronics and Regional Development in Britain

We now turn to a different scale of analysis, concerning the nature and performance of the industry in Britain and its regions. While different considerations come into view at this level it is essential to bear in mind that we are dealing with the *same* structures and processes as were discussed in part II. The international situation is not simply a backcloth to events played out on the national and regional stage; rather the two levels stand in a dialectical relationship of whole to part in which the parts (countries, regions and individual firms and plants) actively constitute, reproduce and transform the whole while simultaneously being influenced by their places within the whole. For example, in part II we made several references to the importance of the nature of the home bases of multinationals, together with the character of host countries and regions, in influencing the development of the international division of labour. We can now examine this more closely in the British case, keeping in mind the international context already illuminated.

The shift down in scale does not merely enlarge the magnification of our analysis so that we see the same things in more detail, for different interests operate at different scales. What we now encounter more clearly are the relationships between the firms and the people of the regions and countries in which they locate. Although, as we shall show, firms and workers in localities have to adapt to one another, their interests are quite asymmetric, involving both interdependence and contradiction. In order to capture the richness of these interrelationships and conflicts of interest and show how they work out in practice we have to go beyond characterizations of the situation in abstract categories. Once again, therefore, we argue the need for a more detailed analysis of the specificities of the industry than is customary so as to avoid the usual oscillation between abstract representations and stereotyped versions of concrete developments.

The following differs not only in these respects from many radical accounts of industrial and regional development but from traditional regional analysis. In the latter the relationship between industry and regional development was generally considered to be adequately covered by multiplier analysis and cognate concerns regarding the economic impact of firms, the firms themselves being treated as 'black boxes'. On two counts we find this unsatisfactory: first, it ignores the social and political aspects of firms' relationships with their regions and how these in turn connect to the economic relationships, and secondly – and particularly in the context of the recession and the internationalized nature of our industry – we are concerned not only with the effects of a given number of jobs but with the quality of those jobs and for how long they are likely to be 'given'. In an era of jobless growth, merely counting enterprises and jobs or assuming that investment entails expanded employment simply will not do.

In part III, then, we aim firstly to present an analysis of the condition of Britain's electronics industry and secondly, and at greater length, to examine the interaction between industry and regional development. After a brief overview of the role of the regions in the industry we shall report in depth on its performance in two radically contrasting regions – South Wales and the M4 corridor. In both these cases we shall pay special attention to a matter of crucial importance to the competitive performance of firms and the experience of work – the changing nature of management–labour relations. Part III ends with a return to the level of the nation state and an analysis of the implications of the neoliberal regime for the industry.

8

The Electronics Industry and Uneven Development in Britain

At the centre of Britain's hopes for an industrial renaissance stands the electronics industry. Compared with the rest of the manufacturing sector the electronics industry has enjoyed strong output growth in recent years: it even managed to avoid the worst excesses of the industrial climacteric between 1979 and 1981 when total manufacturing output slumped. Against this background it is perhaps understandable that the electronics industry has earned a reputation for being relatively resistant to slump and recession. Like steel in an earlier age, it has become a virility symbol for national and regional development agencies. Yet the weakness of Britain's manufacturing industry is well known; even allowing for recession its performance relative to other advanced capitalist countries has been feeble. We saw in part II how Europe lagged behind the USA and Japan in most departments and how it has been colonized by the lower and middle reaches of the corporate structures of foreign multinationals. In turning now to Britain we have to keep in mind not only this context but the general weakness of British manufacturing.

Our aims in this chapter are twofold: first, to examine the structure and performance of the electronics industry in Britain, together with the characteristics of its workforce and, secondly, to examine its role in the structuring of the British space economy. This second part also provides an introduction to the ensuing detailed studies of South Wales and the M4 corridor.

The structure and performance of the electronics industry

The electronics industry in Britain has been subjected to enormous pressures over the past decade and a half, largely as a result of increased international competition and the associated shifts in technology, markets and state regulatory policies which we identified in part II. These pressures have borne most heavily on the indigenous industry i.e. the British-owned electronics industry as opposed to the domesticated overseas-owned sector. This is an important distinction because these two sectors have had different trajectories as a result of an enforced division of labour between them, as we shall see presently. With the growing influx of foreign inward investment it is becoming more and more difficult to speak of a 'British' industry in subsectors such as consumer electronics and semiconductors, both of which are now dominated by British-based overseas firms. Although it is fashionable, at least in official circles, to treat the distinction as having little or no practical value, we argue in chapter 13 that such a view is myopic because it carries grave longer-term risks. We shall therefore examine the electronics industry not only in terms of

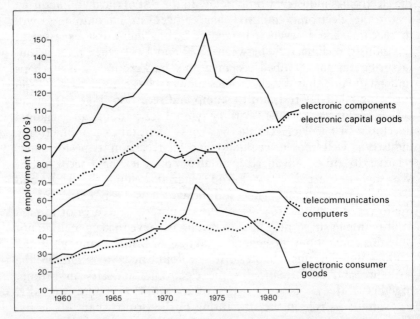

Figure 8.1 Employment in the electronics industry in Great Britain, 1959-83 (based on Department of Employment data)

sectoral changes in output, employment, investment and trade but in terms of the differing roles and characters of indigenous and foreign-owned firms. Our comments on sectoral changes will be brief as the broader processes have already been sketched in part II. However, one sector which we did not isolate at the international level – military electronics – will require closer attention, given its importance in the British industry. Finally in this section we shall look at the changing occupational and gender composition of employment.

Output–employment relationships

While output in electronics has risen rapidly throughout the postwar period employment slumped after 1974 and the industry entered a period of jobless growth from which it is only now beginning to recover. Figure 8.1 illustrates the employment changes for the various sectors from 1959 to 1983. Employment statistics for the period since 1983 are differently categorized but show stagnant or falling employment in all sectors except electronic data-processing equipment.

In terms of output, electronic data processing, followed by capital goods and telecommunications, have taken the biggest share in recent years. However, while output has grown by 180 per cent between 1978 and 1984 in the first of these, this sector also records the highest trade deficit within the industry after consumer electronics, which means that a sizeable part of this recent growth has been met by imports. At the same time employment grew only marginally, reflecting both this import penetration and changing product and process technology (figure 8.2).[1]

In telecommunications, the third largest sector in output terms, two main developments lie behind the shifts in employment. Firstly there have been major shifts in product and process technology. These include the shift from electromechanical and semi-electronic exchanges to fully electronic exchanges (e.g. System X) and the conspicuous growth of the market for private exchanges (PABXs) and associated information technology (IT) products. While such changes are widespread in advanced countries it is worth noting the belated character of the first of these shifts, with pre-digital exchanges still being manufactured in the eighties. British companies have traditionally been cosseted as regards this market for public exchange equipment because of its protected status, while the market for private exchanges and peripheral products, where indigenous firms are much less prominent, has been far more competitive and international in character. Secondly, as we have already indicated, the liberalization of telecommunications in the UK has created a more competitive environment for British firms and has attracted more powerful foreign firms, one of the net effects being that the traditional telecommunications trade surplus has turned into a deficit.

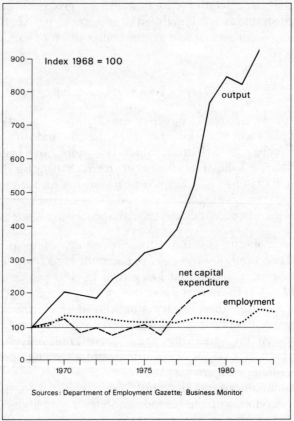

Figure 8.2 Output, capital expenditure and employment in the electronic computer industry, 1968–83 (based on data from *Department of Employment Gazette* and *Business Monitor*)

The position of the indigenous consumer electronics industry deteriorated rapidly during the 1970s largely because of the superior product, process and marketing performance of Japanese companies. Domestic producers were unable to meet the demand induced by the 'Barber boom' (the trade deficit increased by over 100 per dent between 1972 and 1973) thus preparing the ground for an influx of Japanese imports and inward investment (Sciberras, 1980).

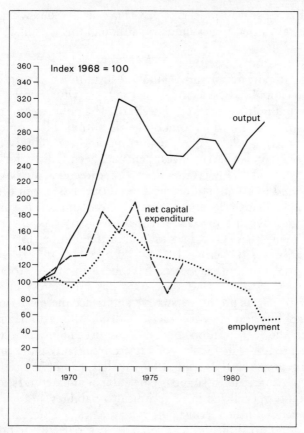

Figure 8.3 Output, capital expenditure and employment in the consumer electronics industry, 1968–83 (based on data from *Department of Employment Gazette* and *Business Monitor*)

Employment in the subsector fell from 69,000 in 1973 to 25,000 in 1981 leading to additional job losses in the component sector as well (figure 8.3). Although the component sector has fared better in terms of output its trade deficit has grown rapidly, particularly in advanced products such as integrated circuits (ICs), and this, coupled with major increases in output per worker stemming from technological change, accounts for its pronounced jobless growth.

The electronic capital goods sector, which covers radio, radar, medical systems and other professional equipment, has in recent years grown more slowly than the electronics market as a whole, 55 per cent as against 60 per cent between 1978 and 1984. One of the distinguishing features of this sector is that it has consistently recorded a positive trade balance, making it somewhat unique within Britain's electronics industry, although this is unlikely to last.

Military electronics

The most important factor behind the traditional strength of the electronic capital goods sector is that it is heavily oriented towards the protected military market; indeed, the UK public sector accounts for some 65 per cent of its total domestic output (Electronic Capital Equipment EDC, 1983). While export markets for military products have grown, foreign direct investment has been limited for obvious strategic reasons. However, under the direction of NATO, an 'administered international division of labour' has begun to emerge. Within this, British locations offer a comparative advantage in R&D largely because the UK aerospace industry has lower salaries than in many NATO countries and a reputation for flexibility and inventiveness (Lovering, 1985).

Substantial parts of the sector have been insulated from market forces or the law of value. The state has an interest in ensuring that its major arms producers survive and hence there is a 'follow-on imperative' in granting contracts; indeed, this is an important factor in the constitution of the 'military industrial complex'. Consequently, bureaucractic contacts are frequently more important than marketing skills (Lovering, 1985; Willett, 1986). As a result of the tendency to continue to support particular manufacturers, each has become quite specialized in its technical capabilites and is often the only indigenous firm serving particular niches. For example, when Ferranti got into difficulties in 1975 it was partially nationalized, since its role as a supplier of communication and control systems to the Ministry of Defence (MoD) was unique and no other British electronics firm was willing or able to take over its role.[2] Moreover, despite the fact that the British government has tried to make arms contracting more competitive, about 60 per cent of contracts are still made on a cost-plus basis – a state of affairs which has brought many accusations of feather-bedding and inefficiency, though estimated permissible profit has been reduced from 20 per cent to 15 per cent (R.P. Smith, 1985; Kaldor, 1982).

Cost inflation is endemic in the industry; Smith estimates that on average each new generation of products costs four times as much as the one it replaces. But it is not merely the method of contracting which causes this. By the very nature of the industry successive demands for products invariably require enhanced performance and price considerations come a very poor second, so whereas technological change in civilian products usually lowers costs (through process

innovation) the emphasis on product innovation in arms production usually raises costs. Often this tendency is amplified because
increased costs mean that fewer units can be afforded, and therefore
production runs are shorter and learning curve and scale economies
are reduced, in turn raising costs further, and so on (R.P. Smith,
1985). Project and small batch production therefore predominate, with
elements of an 'artisanal' labour process, and process engineering is
weakly developed in comparison with industries operating in more
competitive markets. Projects are often long, some of them running
for over 20 years, and these are especially insulated from market
pressures. In one firm producing aircraft navigation systems, only
1.5 systems per year were produced on a 15-year project. At the
end of the development period the amount of business tends to be
limited. This is especially so in Britain because of its small market
(relative to the USA), and the MoD's penchant for overspecifying
products and then choosing simpler and cheaper foreign-made ones
at the end of the development period. There is however a growing
export trade which, in certain products, allows scope for volume
production and is of course exposed to greater price competition.

The extent to which the leading UK firms are defence dependent
is notoriously difficult to define with any precision. After aerospace,
the electronics industry is the largest beneficiary of the MoD's
defence equipment budget, a budget which amounted to $8.2b
in 1986–7, having increased by 40 per cent in real terms since
1979 (Ministry of Defence, 1986a). (In any case much of aerospace
is *de facto* electronics.) In addition to higher absolute levels of
defence expenditure, the share devoted to the electronics industry
has also been increasing – as a result of the MoD's growing need
for 'smart' weaponry – from £941m in 1979 to nearly £2b in 1985
(Ministry of Defence, 1986b).[3] However, there are now signs that
the defence market will not continue to be such a safe haven. For
one thing, MoD expenditure is forecast to decline slightly beyond
1987 after the major increases of recent years and secondly, despite
the much heralded trade performance of the UK military industry,
the UK's share of the world arms market declined from 7.4
per cent to 4.3 per cent between 1963 and 1983 (Kaldor et al., 1986).

Many firms appreciate that the cosy relationship with the government is not entirely in their interest, for it discourages them from
attempting to enter the riskier but often larger and more lucrative
civilian markets; indeed, some found themselves continually diverted
from diversifying in this direction by pressing demands from the MoD
to undertake projects. In any case, the radical differences noted above
between military and civilian work constitute formidable barriers to

transfer. It may seem remarkable that the UK should have a poor record in commercial spin-off from military work when these activities
coexist within the same firms. However, one of the reasons for this
is of wider significance in explaining the peculiarities of British electronics – industrial structure and the internal organization of firms.
The UK industry is dominated by weakly integrated multi-divisional
firms and their defence and civil divisions often have little to do with
one another, hence the low level of lateral technology transfer within
the firm. As a result it is not uncommon to find extremely advanced
capabilities in some divisions coexisting with technical poverty in
another (Maddock, 1982). If GEC is anything to go by, the advanced
capability divisions tend to be those associated with defence activities,
where 32 per cent of employees are scientists and engineers, as
opposed to its consumer products division where a mere 2 per
cent are so qualified (Kaldor et al., 1986). Meanwhile, outside the
large oligopolists few small firms play any significant independent
role at the leading edge. As a result, despite attempts to escape from
this situation the UK's indigenous civil electronics platform 'has been
subsiding, leaving the defence "peaks" standing ever higher relative
to the national electronic engineering plateau' (Maddock, 1983, p.8).

Returning to the British electronics industry as a whole, then,
one of the striking things about Britain, in comparison with the
USA, is the extent to which its industrial structure has remained
stable throughout 'the gales of creative destruction' that have swept
the industry in the past 20 years; Britain has no equivalent of Digital
or Apple. This reflects the lack of involvement of the British firms
in new technologies, the small size of the British market, the limited
mobility of professional labour (and hence diffusion of know-how)
and the role which the British state has played in cementing, indeed
ossifying, the traditional oligopoly: public procurement, public R&D
subsidies and public export assistance have all been massively
concentrated on the traditional 'national champions' irrespective
of their performance (Ergas, 1985; Morgan, 1987).

For an industry which is highly internationalized, most of the
indigenous UK firms are peculiarly dependent upon the UK market.
Unable or unwilling to develop world market strategies, the majority
have been denied the potential benefits of economies of scale and
have remained stretched across too many technologies and product
markets. Not surprisingly, few have managed to lever themselves
into leading international positions in their main product lines.

While there are of course individual success stories – Racal in
data communications, STC in long-distance transmission and Logica
in software systems – these firms remain small by international

standards. At one time some of the greatest hopes were pinned on a UK renaissance pioneered by small firms in the computer and software sectors. Perhaps the most celebrated of all these small firms was Sinclair, which did much to popularize microcomputers in the UK and, as noted earlier, was a key exemplar in the British ideology of high tech. What is most instructive about this firm is that its claims to being a new form of 'ideas-based' company came to grief partly because of some very old British problems, such as poor quality products and late deliveries. Somewhat contemptuous of manufacturing, Sinclair appears to have paid too little regard to the production of its 'ideas', relying instead on subcontractors for this purpose.

The UK software sector is no more immune to these pressures and constraints of the British situation. Although this is a sector in which the UK has a number of strong points, and where niche markets provide some respite from international competition, the cost and pace of technical change are increasing to such an extent that the UK's position appears to be threatened here as well. In the absence of a concerted state-led response it is being squeezed out of volume markets, while the focus of state support means that there is little hope that the UK software industry can avoid becoming a 'defence-only' industry (Advisory Council on Applied Research and Development, 1986, p.15).

Given these trends, it is hardly surprising that dire warnings should have surfaced about the relative decline of the UK electronics industry or, more seriously, about an impending crisis in the UK IT sector, which covers computing, telecommunications and office equipment (Electronics EDC, 1982; Information Technology EDC, 1984). However, these alarm bells are ringing for the indigenous sector as it has retreated into protected public sector markets, leaving the fast growing consumer, commercial and industrial markets to the overseas-owned sector. There has therefore arisen an enforced division of labour between the indigenous and foreign sectors. This is far from a position of equilibrium or equality, however, for foreign firms are, if anything, increasing their grip on the UK electronics market both directly through their UK-based operations and indirectly through imports, so much so that on these trends the UK can soon expect to have not a broad-based IT industry but 'a mixture of inward investment, UK-owned companies employing licensed technology and specialised niche and applications companies' (Information Technology EDC, 1984, p.9). Yet, as we shall see later, the trend towards overseas domination is not perceived as a problem at all in some circles; in the neoliberal scenario, for instance, greater inward investment is seen more as a solution than a problem.

Looking at the overseas-owned manufacturing sector as a whole, foreign firms increased their presence in the decade 1971–81 on a number of fronts: in terms of UK manufacturing employment it was up from 10.3 to 14.9 per cent; net output was up from 13.3 to 18.6 per cent; and net capital expenditure was up from 17.1 to 25.5 per cent (Invest in Britain Bureau, 1984). The UK remains the most favoured European location for US direct investment (over 1,500 manufacturing units in the UK) and, with West Germany, it is the jointly favoured European location for Japanese manufacturing investment (35 units in 1984).

As always, the balance sheet of positive and negative effects of inward investment on the host economy is complicated. The USA and Japan accounted for the bulk of the UK's deficit in IT in 1985, with the USA alone responsible for over 40 per cent of imports against only 10 per cent of UK export markets (IT 86 Committee, 1986). This raises the potentially damning paradox that inward investment may be accentuating, rather than reducing, the UK's trade deficit in IT products because many multinationals appear to operate a deficit on their intra-corporate imports and exports. While the effect of inward investment on output and employment superficially appears to be positive, especially in depressed regions, it does not necessarily produce a net addition in the UK because there are inevitably some displacement effects within the indigenous sector. On the other hand, while foreign firms have often replaced indigenous suppliers with their own foreign-based sources, many foreign firms have greatly improved the quality consciousness of the UK suppliers whom they have engaged. There is also a degree of technological diffusion into the indigenous sector, although this is more often proclaimed than demonstrated.

Besides dominating the fastest-growing sectors, the foreign firms are not only more market orientated and more technologically advanced. They also tend to be more socially innovative with respect to management–labour relations. As we shall argue later, this has bestowed a real comparative advantage on inward investors over their more hidebound UK rivals, though equally it has often been seen as helping indigenous firms by setting an example in these matters.

There are several reasons why the UK is one of the most favoured European locations for US and Japanese direct investment.

1 The UK market is perceived to be among the most liberal in Western Europe and this makes it an attractive bridgehead into the larger EEC market, particularly for US firms because of the 'common culture' of language etc.

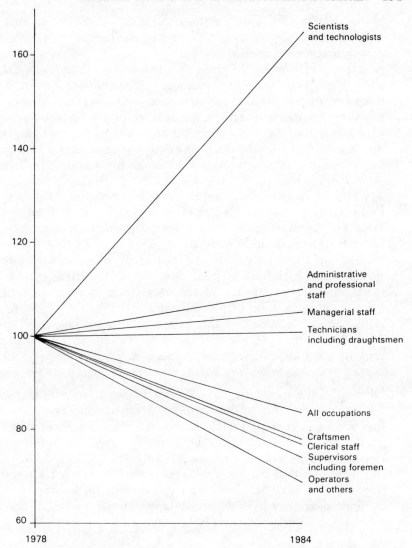

Figure 8.4 Employment by occupation in electronics, 1978 and 1984 (1978 = 100) (based on statutory returns of the Engineering Industry Training Board: from *Manpower in the electronics industry*, EITB, RM85 01, March 1985)

2 Many multinationals believe it is easier to dislodge indigenous producers in the UK than in France and West Germany, where the 'industrial patriotism' of the state is more pronounced.
3 UK labour costs are among the lowest in the major EEC countries, not just for semi-skilled grades but also for technical and

professional grades. This is a tremendous attraction in a country which is still perceived to have 'immense intellectual talent' in its academic community (IBM, 1986).

4 A factor which is rarely given the attention it deserves is the UK's attraction as an international telecommunications hub. Long before privatization the UK sought to offer the cheapest telecommunications charges in the EEC for major business users, particularly on the busy transatlantic route. This encourages multinationals to locate their information-intensive activities in the UK rather than in West Germany, for example, where such charges are far more expensive (Morgan, 1987).

5 The rich pickings in government research establishments appear to be an important attraction in their own right and foreign firms are far more alive to this untapped potential than their UK counterparts (Maddock, 1982);

6 Finally, of course, there is the attraction of generous state subsidies although this in itself does not differentiate the UK from other EEC countries. Nevertheless, foreign firms account for a large proportion of aid offered under the selective financial assistance schemes of the Industrial Development Act: in 1981–2, for example, the foreign sector accounted for 47 per cent of total offers made under Section 7 and 58 per cent of total offers under Section 8 (Invest in Britain Bureau, 1984).

Obviously, the particular mix of attractions varies from sector to sector, from firm to firm and for different functions within the firm. For example, while low wage rates for semi-skilled grades are an important consideration for Japanese consumer electronics firms, this assumes less importance for the likes of IBM, Hewlett-Packard and DEC. For these firms it is far more important to have access to high grade skills in software and computing and such professional labour tends to be cheaper in the UK than in other major EEC countries.

Characteristics of the workforce

The aggregate picture of net job loss (figure 8.1) that we sketched earlier conceals profound changes in the composition of employment along both occupational and gender lines. Firstly, as figure 8.4 shows, higher-grade occupations have recorded net employment growth, especially in the case of scientists and technologists, while lower-grade occupations have been declining, dramatically so in the case of operators. Crude notions of deskilling do not seem appropriate to capture this process; in the higher grades there is reskilling while it is simple reductions in operator jobs rather than deskilling which is dominant in the lower grades. The skills of those who remain in employment in the industry are therefore becoming more polarized and the resulting 'skill chasm' between workers makes it all the more difficult to redeploy those in lower-grade occupations.

As in other industries there is a striking gender division between better-paid, higher-status and more interesting jobs occupied by men and low-paid, low-status and boring jobs occupied by women. Both may involve working with sophisticated equipment but the level of interest and individual control is markedly different. Even where certain jobs appear to be done both by men and women such as wiring, closer examination invariably shows the particular kind of work done by the men to be more interesting and to involve greater control. The managerial, scientific and technical occupations are male bastions which show little sign of feminization despite severe labour shortages in certain jobs. The male dominance of managerial positions is overwhelming; for example, in the electrical capital goods sector, only 67 of the 5,000+ managers were female (Engineering Industry Board, 1985). However, discrimination operates powerfully outside the workplace and managers often bemoan the lack of applications for engineering jobs from women. The assumption in the business literature that all managers and key employees are men is universal and not regarded as worthy of comment, let alone justification.[4] While sexist job descriptions are not allowed by law, they are common in informal discourse. For example, one firm described its occupational mix in the following terms: '4 Directors, 7 management/engineers, 5 technicians, 1 admin. girl, 1 receptionist, balance – 42 assembly girls' (Source: interview). This concentration of women in the operator and clerical grades is typical. The reasons for management's preference for women for this kind of work are well known; lower wages, significantly higher productivity in operator jobs than men (even allowing for their higher rates of absenteeism) and their lesser involvement in union activity. All these characteristics are of course socially produced and therefore subject to contestation and change.

As the industry expanded in the sixties and early seventies (a time of labour shortage) so the demand for operators and hence women workers grew, only to fall again when their jobs were displaced by a combination of foreign competition and effects of automation and integrated circuitry. In some cases technical change was associated with gender change, usually in ways predictable according to the type of skills; for example, where testing was deskilled it would usually become a 'woman's job'.[5] At the same time, as products became more information-intensive so the demand for managers, technologists and technicians grew – occupations from which women were largely excluded. There has therefore been a 'defeminization' process within the electronics industry since women accounted for over 90 per cent of net job losses between 1970 and 1981 (Engineering Industry Training Board, 1982).[6]

The electronics industry and the British space economy

The contribution of an industry to regional development clearly depends not just on the level of activity in a region but also on the character of that activity. Different activities often correspond to the hierarchical spatial divisions of labour of large multiplant firms, as when 'top-end' activities (such as control and research and development functions) are restricted to the main metropolitan region, while 'bottom-end' activities (such as production) are more widely diffused. Hence it is common to find that the proportion of unskilled or semi-skilled workers is much higher in the peripheral regions than in the main metropolitan region. Even small specialist single-plant firms tend to locate in areas according to similar principles, so that research-intensive small firms, for example, locate in similar areas to similar activities in large mutliplant firms. Clearly, the type of activities through which a region is inserted into the wider national and international divisions of labour will have different effects within the region in terms of income, skill and vulnerability to closure. In the following chapters we shall explore some of these forms of development by looking at the growth of the electronics industry in two radically different regions: South Wales and the M4 corridor, the latter being an important component of the English 'sunbelt'. But first, in order to set this in context, it is necessary to provide a broader sketch of regional differentiation in the electronics industry in Britain.

Underlying the geographical pattern of employment numbers (figure 8.4) lie marked qualitative variations. The aggregate spatial division of labour within the electronics industry in Britain exhibits both hierarchical and sectoral/regional dimensions, with the south-east dominating both. In other words, the south-east not only dominates the regional league as a location for front-end functions but, with some 53 per cent of total UK employment, it also dominates in absolute terms, a reminder that there is still a pronounced sectorally based spatial division of labour.[7] So, for example, while the relative proportion of semi-skilled operators in the electronics capital goods sector (Standard Industrial Classification minimum list heading (MLH) 367) was higher in peripheral regions than in the south-east in 1981 (55 per cent in Wales against 18 per cent in the south-east), the absolute number of operators is much higher in the south-east (10,195) than in regions such as Wales (222). Clearly, then, back-end activities are by no means the preserve of the periphery, contrary to certain misleading stereotypes of the spatial division of labour (Sayer, 1985). Within the south-east it is Greater London and the early post-war industrial centres – such as Letchworth and Hatfield – rather than the outer south-east, where the routine manufacturing activities tend to be located.

In fact, these figures seriously underestimate the south-east's dominance because they are based on a narrow definition of electronics (MLH 363–367)[8] which excludes much of the computer services industry, in particular, the critically important software sector. Indeed, it is in this industry that the south-east's position is strongest of all since it accounts for over 70 per cent of UK employment, with Greater London being by far the most important concentration (Department of Industry, 1982b). The reasons for this heavy concentration in London are twofold: proximity to major clients (such as government, the City and corporate headquarters) and access to scarce professional labour (Greater London Council, 1985).

The predominance of high-level functions in the region differentiates the south-east from the rest of the country. Associated with this functional specialization, which clearly betokens an elite occupational structure, is another distinguishing feature, namely the south-east's unrivalled capacity to spawn and sustain new firms. Generally speaking, the map of professionally skilled high-income social strata is a surrogate for the map of new firm formation. And the scope for new firm formation in the electronics industry, whether in the form of subcontracting or of a more autonomous kind, is immense in comparison with traditional industries like steel, where there is little or no growth potential, where technological change is more pedestrian and where corporate oligopolies dominate the market. As we shall see, the technical character of the industry in the south-east is both cause and consequence of the predominance of professional labour markets in certain parts of the region and, taken together, this constitutes a rich seam from which new, often high-grade, activities are formed.

The roots of the regional pattern lie in the concentration of early electrical industries in and around London (Hall et al., 1987). The relative wealth of the south-east in the inter-war period, together with its early access to the National Grid, made it the major regional market for electrical firms. This early localization of the market appears to have been responsible for the extraordinary concentration of US inward investment in electrical engineering in west London and Slough during the thirties and the early post-war boom period. Later,

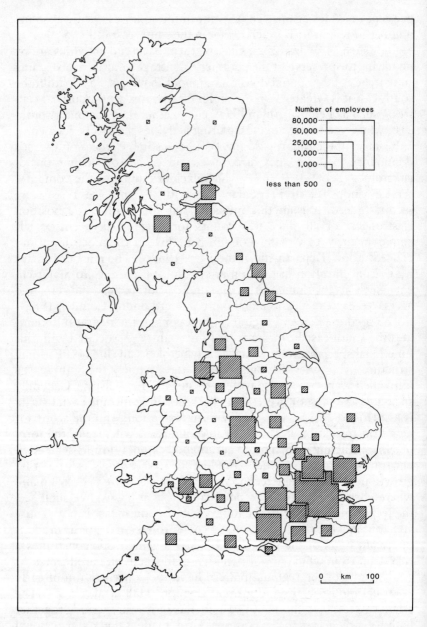

Figure 8.5 Employment in electronic engineering, 1981, by county (based on data supplied by the Department of Employment)

this localization became self-perpetuating since the more scientific industries, such as aerospace, instrument engineering and electronics itself, are major users of electronic capital equipment and this sector has long been the most concentrated in the south-east. Access to markets, suppliers and skilled labour all contributed towards this tremendous concentration of the industry in the metropolitan region.

The south-east has not, however, been immune to the wider shifts in the British space economy, especially as regards the so-called 'urban–rural' shift. Over the past two decades London, along with all other metropolitan areas, has suffered a huge decline in manufacturing employment; in the electronics industry the chief losses have occurred in consumer products and pre-digital telecommunications equipment. These losses were due to a combination of *in situ* loss and decentralization (Massey and Meegan, 1979; Leigh et al., 1983). In the 1960s and early 1970s, a combination of factors – labour shortages, deskilling of the labour process and regional policy – encouraged some decentralization of more standardized work to the peripheral regions and to the outer fringes of the south-east itself (Keeble,1976: Massey and Meegan, 1979). At the same time many front-end activities moved out to the more accessible parts of the south-east, particularly Berkshire, Hampshire and Buckinghamshire, forming part of the crescent or sunbelt of relative prosperity around the western and northern sides of London from Southampton to Cambridge, with outliers in towns such as Swindon and Bristol. For example, Cambridge has developed a notable concentration of new electronics firms, particularly in development and software work, although their employment effects are minor.

Nevertheless, Greater London remains by far the biggest centre of electronics employment in the country. Even after employment declines of 23 per cent between 1971 and 1978 and a further 16 per cent between 1978 and 1981, it is still over twice the size of the much-vaunted Scottish industry. Moreover, its mix of activities remains the most diverse, including both large numbers of small producers of highly specialized equipment (e.g. for television studios) and major manufacturing plants of firms like STC, Plessey and Thorn-EMI. On the other hand, even the favoured 'rest of the south-east' has not escaped job losses in the sector. Between 1978 and 1981, it suffered a net loss of electronics employment, the expansion in Berkshire, Surrey and Kent being overshadowed by losses in Essex, Sussex, Hampshire and Buckinghamshire.[9]

Too often the attractions of the south-east are reduced to its 'environment', as though this were some natural endowment. What needs to be more widely understood, however, is the enormously important role that the state has played in creating the comparative advantages that the south-east enjoys over the rest of the country. First and foremost there is the long-established role of the MoD in the region, perhaps the single most important factor behind the concentration of the electronics industry in the south-east, due to the traditional emphasis placed upon synergy between the MoD and the defence industry. The south-east is far and away the major regional beneficiary of MoD spending measured in terms of the value of MoD contracts (Short, 1981), it is the most militarized zone in terms of army camps and air bases and it is overwhelmingly the home of the military elite (Fotherfill and Vincent, 1985, p. 18). Overall, the general picture has been summarized thus:

> From Stevenage in the north to Crawley in the south there are some 64 companies which have major contracts with the MoD. . .Roughly 40% of all defence expenditure is in the South East. It provides 300,000 civilian jobs in defence and defence-related industry. The South East is therefore something of a defence-orientated economy, locked into the need for constant renewal of contracts for weapons systems and equipment. (Greater London Council, 1985, p. 287)

Again, a hierarchical spatial division of labour is apparent with the south dominating the front-end activities and the more routine and less skilled production being more widely dispersed. But there are other more obvious ways in which the state has contributed to the advantage of the south-east. Here we need only think of central government itself, access to the corridors of political power being an important factor behind the concentration of corporate headquarters in London. Then there are the government research establishments, the majority of which are in the sector of the south-east west of London: these not only formed an early market for electronic equipment but also helped to amass pools of technical labour and greatly enhanced the region's scientific infrastructure. Nor should we forget other forms of unique public infrastructure such as Heathrow Airport, access to which has been a powerful stimulus for much of the growth to the west of London. In each of these ways we can see how great the role of the state has been in fashioning the comparative advantage of the south-east.

Apart from some early centres such as those associated with Ferranti in the north-west and in central Scotland, 'peripheral Britain' has been dominated by routine hardware manufacture. Regions which for most of the last 50 years have had assisted area status gained first from the strategic decentralization of key industries such as telecommunications since the war. Further invest-ment in these areas came particularly during the sixties and early seventies when, prior to the recession and the displacement of many manual tasks by automated assembly and testing equipment, firms increased their demand for cheap labour – usually mainly female – and enjoyed generous employment and investment subsidies in the development areas (Massey and Meegan, 1979; Morgan, 1985, 1986).

These areas too were hit by recession and job displacement in the industry in the seventies and eighties, but while some, such as Merseyside, suffered major employment losses others with better reputations regarding industrial relations and more support, or more favourable mixes of electronics subsectors, were able to offset this partly or wholly by attracting inward investors.[10] While such inward investment still tends to be in back-end activities only, the technology is still relatively advanced in British terms. Later, in chapters 9 and 10, we shall look at developments in one of these peripheral regions – South Wales – in more detail, contrasting it with part of the south-east, but before moving on to another part of the British periphery requires more attention.

In part II we warned against seeing the international spatial division of labour in stereotyped and fixed forms and noted how the newly industrializing countries were changing their position. The same applies at subnational levels and within Britain's outer periphery the position of central Scotland, popularly known as 'Silicon Glen', is especially noteworthy in this context. While it has a very high degree of external control it diverges significantly from the peripheral branch plant economy stereotype, its status within the industry lying midway between Wales or Ireland and the south-east.

Central Scotland's electronics industry has been perceived as one of the great regional success stories. Having once been dominated by the traditional industries of coal, steel and shipbuilding, Scotland is often taken to be an example of what other older industrial regions might also be able to achieve. This view is encouraged by the fact that the growth of the electronics industry in Scotland is an example of (conscious) state-sponsored development stretching back to the exigencies of a war economy when Ferranti, a leading military contractor, moved into Scotland for security reasons (Firn and Roberts, 1984). Since then a battery of regional policy incentives, in conjunction with a well-connnected 'regional' state and an innovative development agency, have been pressed into the tasks of attracting inward investment and building up a supportive technical infrastructure. Apart from Wales, no other standard region in the UK comes remotely close to having such an institutional arsenal at the regional level. This Scottish state apparatus was already aware of the potential of electronics-based regional development well before this, however, having used Ferranti as a conduit through which MoD work could be shared among Scottish-based firms in the 1940s. Furthermore while other development areas addressed their overtures to the Board of Trade, then responsible for the execution of regional policy, the Scottish establishment realized that the science-based industries of the future were being shaped by military contracts – hence its strenuous efforts to develop local subcontracting expertise in Scotland (Burns and Stalker, 1962). Although Ferranti is today the largest company in the electronics industry in Scotland, accounting for nearly 20 per cent of total employment, it has not had any great impact in spawning new spin-off companies in the region nor has it developed the local linkages one might have expected given its longevity in the regional economy. This has been attributed to the secretive nature of its military work and to its policy of maximizing in-house production. Nevertheless, Ferranti did contribute to the regional economy by recruiting key engineering skills which were later to help attract foreign entrants (Firn and Roberts, 1984).

The main impetus behind the growth of the electronics industry in Scotland has come through inward investment from large English and overseas companies. In terms of employment, the major sectors are now electronic data-processing equipment, defence systems and electronic components. In 1985 total employment was estimated at 43,800, up by 4,500 since 1978, an increase accounted for entirely by US-owned firms, most of which were using Scotland as a bridgehead for their pan-European strategies.[11] However, while employment levels in electronics are now higher than in any of the traditional sectors of shipbuilding, steel or coal, they still only amount to 6–10 per cent of the Scottish workforce and employment growth has now stagnated despite the numbers of new entrants. Such is the dominance of large plants that a mere 5 per cent of all units account for some 50 per cent of all employment. Overall, the overseas-owned sector accounted for 48 per cent of employment in the industry in 1985 and 42 per cent of this is US owned. There are signs that the area now has sufficient numbers of branch plants to support some local suppliers where formerly most supplies and services were imported. However, while this suggests that the 'cathedrals in the desert' syndrome may be weakening, the much hoped for growth of indigenous firms has not emerged on any sizeable scale and by 1983 employment in the Scottish-owned sector had reached only 16 per cent of the total (Firn and Roberts, 1984; Industry Department for Scotland, 1986).

Nevertheless, the operations of the US firms in Scotland include a good deal of development work, albeit of a low level mainly involving adapting products to the nuances of European markets.

But while such work is of a secondary status compared with that of their parent companies, it can be quite advanced by the standards of the civilian side of the indigenous UK firms. In other words R&D itself needs to be unpacked because these activities do not carry the same innovative potential in Scotland as they do in California, for example. And yet Scotland is not without indigenous technological strengths, although these are to be found not so much in the indigenous industry as in the university sector, where an internationally respected expertise has been developed in such critical technologies as very large scale integration (VLSI), opto-electronics and artificial intelligence (Henderson, 1986). As a result the attractions of Scotland as a location for inward investors should not be thought to consist only of regional grants, important as these have been in the past (Scottish Development Agency, 1982). This growing pool of skilled labour gives Scotland at least one major advantage over the agglomerations in the English sunbelt, namely that, once recruited, key personnel are more easily retained because distance confers some protection against the debilitating poaching practices that are so rife south of the border. Further, an attraction which Scotland enjoys over peripheral regions like South Wales in the eyes of multinationals is the fact that a large non-unionized enclave has developed in the Scottish-based electronics industry. Pioneered by US firms like IBM, non-union practices have been aped by a host of later entrants.

In its attempts to up-grade the Scottish-based electronics industry the Scottish Development Agency is effectively trying to modify the parameters in which this industry has to operate. Its scope for success may be extremely limited but we should not ignore the actual potential for change even at this regional level; as we saw in part II, to do so implies that inherited structures are somehow immutable and impervious to state-sponsored initiatives.

The electronics industry has undoubtedly produced local agglomerations of growth, and these are all the more conspicuous in mature regions like South Wales and central Scotland. Too often though these islands of growth are judged in and for themselves, the result of a dangerous parochialism. It is dangerous precisely because it obscures a wider reality, namely that the indigenous UK electronics industry is gripped by relative decline and that regions like South Wales and central Scotland are still among the most depressed in the UK. It may even be the case that the very growth of a local or regional agglomeration occurs at the expense of, rather than in addition to, employment and output in the national industry, which is what appears to have happened in the case of US semiconductor firms in Scotland and Japanese consumer electronics firms in Wales.

Hence the regional dimension is an important one because, now that regional coalitions have emerged to advance the claims of their particular 'turf', a national industrial strategy could well fall foul of territorially based conflicts of interest. Nowhere is this more evident than in the perennial regional, and indeed local, battles to 'capture' internationally mobile investment in the electronics industry. While the local benefits of this 'game' may be tangible enough, the costs, though diffuse and therefore less palpable, can be every bit as real, though these are experienced elsewhere of course.

Introducing South Wales and the M4 corridor

If the branch plant stereotype needs significant revision in the Scottish case, its essentials are far more visible with respect to South Wales. At the other extreme, the English M4 corridor is particularly associated with the front-end activities of the industry. These two areas lie at opposite ends of dominant hierarchical spatial divisions of labour and represent some of the biggest contrasts in Britain's social and economic geography. Electronics firms operate in both areas, but they tend to be of radically different kinds. Yet, as we saw at the international level, difference need not mean independence and there are many links between the differing kinds of activity carried out in the two areas, some of them internalized within the same firms.[12] There are also strong interdependencies between these activities and the character of the areas in which they are set. It is to these contrasting regional contexts that we now turn.

Regional legacies

Past forms of industrialization can exert profound effects on a region long after the demise of its formerly dominant sectors and this structural legacy may constrain or facilitate the emergence of new, more advanced, sectors. This emergence may come about by three main routes: new firm formation, diversification of existing firms and inward investment. Of these, the first is limited by the social composition of the region, a sizeable managerial class being a necessary (but not sufficient) condition for new firm formation on any significant scale. The second and third routes are strongly influenced by the diversity and malleability of the traditional practices of the region's labour force. The significant characteristics of labour culture from capital's point of view go far beyond wage norms to received attitudes to work, types and extent of unionization, gender composition and racial mix. In the case of expansion involving large contingents of higher-skilled and managerial labour, the intangible perceived qualities of the landscape and social mix of the locality tend to affect the three routes to expansion, and of course these and many other features vary enormously between different regions.

South Wales

Some measure of the former dominance of the traditional industries of coal and steel lies in the fact that, together, they employed 76 per cent of the insured working population of South Wales at the time of the 1930 Census of Production. Few parallels exist, outside peripheral countries, for such a commitment of capital and labour to so narrow a sectoral base, and it was not until after 1945 that the regional labour force showed signs of becoming sectorally – though not occupationally – diversified. This restricted form of industrialization, economically buoyant in its extreme form only until 1921, wrought such durable effects that many can still be discerned today. Among the most important we would cite the following:

1 The class composition of the region was overwhelmingly prolet-arian and, while this stark social profile has attenuated since the 1930s, South Wales remains the nearest approximation to a single (working) class region in Britain;

2 With few exceptions, external control of the region's coal and steel industries was already an established phenomenon by the later inter-war years and the consequent absence of an indigenous 'business class' helps to explain the dearth of alternative industries prior to the Second World War and the limited potential for indigenously based regional development thereafter;

3 The nature of work in the coal and steel industries allowed workers a considerable degree of autonomy over their day-to-day work tasks and the absence of a labour market culture tutored in 'working to the hooter' – an official euphemism for the discipline associated with factory regimes like Ford's – continued to be remarked upon even in the 1960s;

4 The formerly dominant industries fashioned a profoundly dichoto-mized gender division of labour in which women were largely excluded from wage–labour relations. The legacy of this division is still apparent: despite the waged labour opportunities that emerged after 1945, principally in the service sector but also in light industries such as electrical engineering, Welsh female activity rates remain among the lowest in Britain. When firms

have sought new production locations needing new sources of operator labour, such untapped reserves of female labour have proved a significant location factor (Massey and Meegan, 1979; Massey, 1984). More recently, with the post-war contraction of the coal and steel industries and the enlargement of the public services sector, the gender recomposition of the employed Welsh working class has been one of the most conspicuous features of the last two decades (Cooke, 1981; Winckler, 1985);

5 The highly unionized character of the coal and steel industries lies at the root of the region's traditional labourist culture, a culture whose persistence is shown in the fact that the newer service and manufacturing sectors are also relatively highly unionized. South Wales remains one of the major and most consistent bases of the Labour Party. Its reputation as a radical region was enhanced by its record during the 1984–5 miners' strike, in which the South Wales coalfield showed by far the highest and most tenacious support for the strike (Cooke, 1984).

This then provides a sketch of the context facing new industries such as electrical engineering in South Wales. While it is true that regional assistance and South Wales's accessibility to south-east England relative to that of other assisted areas have played their part in encouraging new industry, these are by no means the only factors. No firm of any standing can afford to ignore the social context in which it is to operate, for, although regional aid might give a welcome short-run financial boost, the long-term productivity, and hence profitability, of its operations depends heavily on how they can adapt to this context.

The M4 corridor

A predominantly rural area, unblighted by heavy industry, the M4 corridor differs strongly from South Wales both visually and in socio-economic terms. It includes many small, rapidly growing towns, some of which, such as Wokingham in Berkshire and Fleet in Hampshire, have been identified amongst the most prosperous in Britain in the eighties: this is stereotypically Thatcher's Britain. As a more mixed area in terms of class, with above average proportions of owners of capital and managers, the corridor offers far more opportunities than South Wales for new firm formation aned industrial diversification. On the other hand, the lack of concentrations of highly dependent workers as large as those of

South Wales means that the area is generally less suited for intensive shift work. There are, however, unskilled workers in some of the smaller towns of the corridor who, through their greater isolation and lack of accumulated organizational strength or labourist culture, are in a weaker position than their counterparts in Wales.

Despite the rural image of the corridor, its major towns have certainly not escaped early industrialization. The electronics industry is in some parts completely novel, in others an outgrowth of earlier activity. Although the much-noted pool of skilled labour at the eastern end of the corridor is a product as much as a cause of the localization of the electronics industry, the area was considered to be the best for scientific and defence-related labour even in the early fifties, thanks to the presence of government research establishments. Local universities elsewhere were among the leaders in electrical engineering, notably Manchester and Edinburgh, and yet their localities failed to generate a labour pool of comparable size. But the Thames Valley had an additional advantage: even then, the attractions of the environment for science graduates, who would have been even more overwhelmingly middle class and of southern origin than now, appear to have been strong.

In Bristol, a pool of skilled engineering labour had built up around the aerospace industry which was in any case increasing its electronics content. Reading's traditional industries were dominated by beer, bulbs (the horticultural variety) and biscuits, though thousands of jobs have been lost in these as firms have contracted or moved out of the town. Swindon had developed around major railway works – closed in 1985 – and had added a number of manufacturing activities, particularly in automotive engineering, in the post-war period. As an 'expanding town' it was one of the few areas of southern England to be permitted rapid industrial and residential expansion in the 1960s.

Slough, at the eastern extremity of the corridor, differs from the rest of the area in that it was the most affected by the 'new industries' of the 1930s and the immediate post-war period. Among the workers who filled these jobs were thousands from the 'distressed areas'. including 16,000 from Wales alone, brought in under the Government Transference Schemes. Slough's huge trading estate captured a large proportion of the American firms which set up in Britain at the start of the last 'long wave', including some electrical engineering firms like Black and Decker and Sperry. It remains a solidly manual working-class town with a large black population and over a quarter of the workforce in semi-skilled or unskilled jobs. In other words, the corridor offers not one but a variety of environments for capital.

Table 8.1 South Wales and the M4 corridor: statistical indicators

	Unemployment (Oct 1985) (%)	Average gross weekly earnings (full-time, April 1985)		Cars per 1000 of population	Housholds lacking inside W.C., 1981 (%)	Women (15–65) earning wages (%)
		Males on adult rates	Females on adult rates			
South Wales						
Gwent	17.2	175.5	116.6	293	4.3	56
Mid Glamorgan	19.0	175.2	119.5	227	9.2	55
South Glamorgan	14.4	187.8	123.1	289	3.4	59
West Glamorgan	16.2	181.6	119.1	272	5.0	57
M4 corridor						
Avon	11.4	192.3	124.4	342	1.5	63
Berkshire	7.3	212.2	138.8	404	1.7	60
Hampshire	9.9	196.4	126.2	345	2.0	60
Wiltshire	9.9	181.1	123.1	367	1.8	61

Source: Central Statistical Office, *Regional Trends* 1956

The development of these centres, plus some smaller ones like Newbury and Maidenhead, was further aided by congestion within west London and the construction of the M4 motorway and the 125 mph rail service from London to South Wales. These enabled the zone of rapid access to London and the growing labour pool to stretch out westwards,[13] attracting not only electronics firms but a number of decentralizing company headquarters, data-processing centres and 'back offices' (offices concerned with routine work); indeed, many towns in the corridor owe more of their success to the latter activities than to electronics. A further and lasting attraction, particularly for foreign firms, was access to Heathrow's international airport, the busiest in the world.

The implications of these preliminary remarks on the regional legacies of South Wales and the M4 corridor are clear enough. Although capital which is new to an area will attempt to reconstitute a traditional labour force to suit its own specifications, the regional legacy may imprint itself on new employers. In some cases the legacy may be wholly advantageous, but where it is considered too burdensome, alternative strategies might be to tap new sources of labour (e.g. women) within existing locales or to seek out new locales, within the same regions or elsewhere, where the social practices of labour are more permissive. As we shall see, both of these have occurred in South Wales and the M4 corridor.

Statistical contrasts

The contrasting nature of the electronics industry in these two areas owes much to the differences in their economic and social character.[14] In terms of objective characteristics, what comes through most strongly are class differences (table 8.1). South Wales is an area of net out-migration; the M4 corridor, especially Berkshire, is one of net in-migration.[15] South Wales has been hit far harder in the recession, with over 28 per cent of industrial jobs lost between 1978 and 1981, while the south-east (excluding London) lost less than 18 per cent and had more non-manufacturing jobs to compensate.

Politically, the M4 corridor is solidly Tory though in the past Labour has held seats in Bristol, Swindon, Reading and Slough. Mid Glamorgan has held the highest percentage Labour vote in Britain (52.9 per cent, less than 30 per cent Tory), while over 50 per cent voted Tory in Berkshire's and Wiltshire's constituencies.

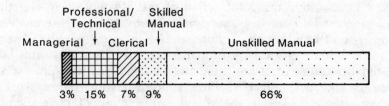

Skill Structure of Jobs in 21 S.Wales Electrical Engineering Firms 1983

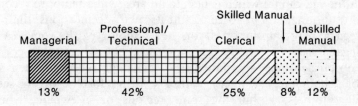

Skill Structure of Jobs in Electronics Firms in Berkshire 1984

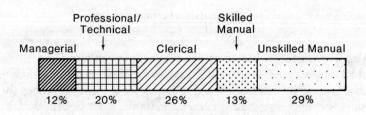

Skill Structure of Jobs in Berkshire Firms 1984

Figure 8.6 South Wales and Berkshire: contrasts in skill structure (based on data from the Engineering Industry Training Board and Royal County of Berkshire, 1985 and 1986)

Wages are not as different as might be expected, thanks to the presence of highly organized manual workers in South Wales and the number of poorly organized rural workers in the corridor.

Perhaps the most telling contrast is to be found in the skill profiles of the two areas. The profile for electrical engineering in South Wales is compared in figure 8.6 with that for electronics firms in Berkshire, the main concentration of the sector in the corridor.[16] As we shall see, this is of considerable significance for the regional development effects of the industry.

Mythologies of Merthyr and Maidenhead

It is of no small importance that the *landscapes* of South Wales and the M4 corridor differ radically in both actuality and popular perception. Both the appearance and the affective associations of the two areas speak of markedly different cultural and class bases. Up to a point it matters not that the popular perceptions contain a large element of myth, for they still influence those who have the wealth to make locational choices, be they those of capitalists regarding the location of capital or those of highly skilled workers regarding the use of their own labour power. Nor does it matter that the usual descriptions of these perceived landscapes are loaded with cliches; on the contrary, cliches are the very stuff of perceived landscapes.

South Wales's gaunt landscape of high moorland divided by deep, once beautiful, valleys lined with tight rows of terrace houses, roads and railways and scarred by mining and industrial decay speaks of the hard labours and poverty of a classic industrial working class, a class whose legendary communal solidarity now seems oddly anachronistic to most English people. There is also the lower-lying coastal plain, more pastoral, but also with its grimy, gloomy industrial towns and works. Like most of peripheral Britain, industrial South Wales is rarely given a thought by the English middle classes except when political events there intrude on the centre.

In radical contrast, there is the lower, more arable landscape of the M4 corridor – a 'fat-cat' country of shallow valleys, rolling downlands, patchwork quilts of fields, punctuated by public schools, market towns and twee villages with stone-walled and thatched-roofed cottages, an area threaded through by the placid waters of the upper Thames with its pleasure boats and banks lined with the gardens of large houses – all redolent of a comfortable, leisured wealth, removed from the industrial working class and yet still accessible to London.[17] Opinions differ on which is the more attractive – the wilder and more rugged countryside of South Wales or the 'tamed rurality' of southern England – but it is the latter, and its proximity to London, which has long appealed to bourgeois tastes at least as an area in which to live. Indeed, of all the landscapes of England, the south has arbitrarily come to be seen as quintessentially English.

The modern image of the M4 corridor grafts onto this: associations of executives (white and male of course) in up-market cars leaving their 'Tudorbethan' Barrett homes to catch a jet from Heathrow or cruise up the M4 to their high-tech offices, windsurfing on the reservoirs of the Thames Valley at weekends and working at home on the micro. The only modern image of South Wales available is that ingeniously constructed by the Welsh Development Agency (WDA) in its advertisements. In its attempt to graft the ideology of high-tech success onto the public image of Wales the only insignia the WDA can afford to acknowledge in the latter is the male voice choir, and this in a setting not of mining communities but of the green, unindustrialized valleys of rural Wales. Again, many aspects of these images are misleading, but this has not reduced their significance.

In the following chapters we shall see how these contrasts – actual and perceived – impact on the industry in terms of company performance, type of work, types of workers, management–labour relations and organization. Once again we should say that although the amount of detail regarding company performance may seem unusual we feel that it is justified; there has been too much rhetoric and uninformed gesturing from all parts of the political spectrum regarding the nature of 'branch plants' and 'high-tech firms' and some demystification is overdue. While some of the details may in themselves be of transient and parochial value, they are still important to the firms in question and they give something of the flavour of the problems confronting the industry. Nevertheless, as we shall see, certain dominant themes emerge.

9

A Modern Industry in a 'Mature' Industrial Region
The Electrical Engineering Industry in South Wales

in dual
company B

With just 23,000 employees in 1981 (over 13,000 in electronics), South Wales occupies a lowly position in the industry in terms of both level and status of employment. Yet it has been proclaimed a 'success' by the West Development Agency (WDA) under the usual superficial label of 'high tech'. To get a better assessment of the strength and position of the industry in South Wales it is necessary to look at the specific nature of the plants and assess them within their competitive context in order to judge their performance.

Electronics in South Wales has been the manufacturing sector most resistant to recession and has consistently achieved an above-average growth of output. While aggregate electronics employment grew by a third between 1971 and 1981, it still constitutes only 3.4 per cent of Britain's employment in order IX of the Standard Industrial Classification (SIC) or 6 per cent of total manufacturing employment in South Wales and is never likely to approach the peak employment levels – and the social significance – attained by coal or steel.

Electrical engineering in South Wales provides striking contrasts with the region's traditional industries of coal and steel, but it also contains some significant internal contrasts. Plants in the sector range from early post-war electrical engineering factories to the manufacturers of advanced electronics products which arrived in the seventies and eighties. The first group is mainly British, e.g. Smith's, Thorn Lighting and GEC Telecommunications, but includes a few foreign firms such as Hoover. The second group consists of foreign firms such as Matsushita, Siliconix and Mitel and includes five of South

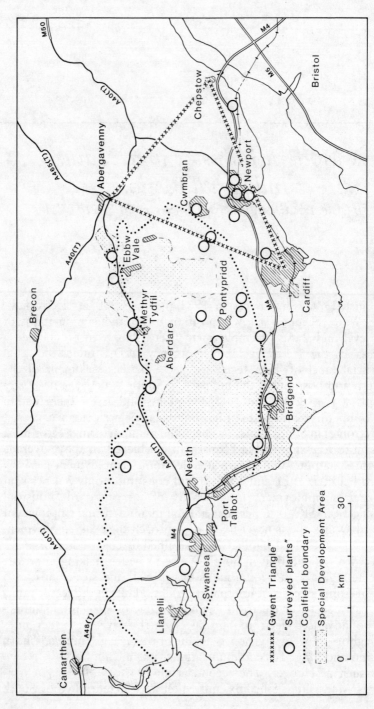

Figure 9.1 *The electrical engineering industry in South Wales*

xxxxxxx "Gwent Triangle"

○ "Surveyed plants"

.... Coalfield boundary

▨▨▨ Special Development Area

0   30

km

Wales's eight Japanese firms (a concentration unparalleled in Europe). Most of these have arrived since 1974. As we shall see, the technical and social contrasts between these two groups are considerable.

Only one firm of any size in the sector actually originated in South Wales and the region has the highest degree of external control of any in Britain in the sector. This is to be expected given the dearth of professional and managerial people in the region and the fact that the industry has been the most prominent in post-war inter-regional movements of plants and among the sectors with the highest penetration of foreign capital. In fact electrical engineering is more dominated by overseas companies, and has had more plant openings, than any other sector in Wales. However, the activities carried out in the Welsh plants tend to be of low status within corporate hierarchical divisions of labour. They are also heavily weighted towards older, more mature and less science-intensive subsectors such as consumer electronics and simple components rather than leading-edge information technology (IT) products.

Regional assistance has certainly influenced the many inward investors. In the early seventies it took the form of a labour subsidy[1] and then later regional development grants for investment regardless of employment effects. Just how important regional aid was in attracting firms is of course unclear given the usual secrecy and *ex post* rationalizations surrounding location decisions. In the case of the Japanese plants, there appears to have been a general desire to comply with central government's wishes in going to an assisted area (no doubt fostered by fear of protectionist exclusion) and a clear imitation effect, supported by special targetting by the WDA (and before it, the Development Corporation of Wales) of Japanese companies. While many other peripheral regions have received the same kind of aid, South Wales perhaps has an advantage over them in being the most accessible to south-east England and the M4 corridor. A few firms have capitalized on this by 'retreating' over the Severn to get grants while escaping from the labour poaching and salary inflation characteristics of the English end of the M4 corridor. The area of east South Wales which the WDA refers to as the 'golden triangle' (figure 9.1) might therefore be dubbed the 'subsidized end of the M4 corridor'. Generally, most new firms have located within a few miles of the M4 motorway though, as we shall see, this is not simply to do with the coalfield's poor road system.

In order to assess the performance of the firms in the area we shall examine them in their causal contexts, rather than simply taxonomically as members of classes having certain common properties, though in some cases these coincide with the taxonomic

Minimum List Headings (MLHs) of the SIC. The 27 plants surveyed in the research constituted over 55 per cent of employment in electrical engineering and over 90 per cent of electronics employment.[2]

Consumer electronics

The experience of South Wales in this sector is radically different from that of Britain as a whole. As we saw in chapters 4 and 8, Britain's consumer electronics sector was hit harder than any other part of the industry by a devastating combination of market stagnation and competition from Japanese firms selling superior, though rarely cheaper, products. The result was that over half the plants owned by established firms (British, Dutch and American owned) were closed and employment fell by two-thirds from 1974 levels. The accompanying invasion of Japanese firms entering Britain to pre-empt British and EEC protectionism saved the 'British' industry by conquering it. Yet, typically, their role in decimating the established indigenous suppliers was quickly forgotten and they were soon welcomed as providers of new jobs. In South Wales at least, which had previously had only one plant of any size in consumer electronics, the arrival in the second half of the seventies of four new Japanese plants actually brought an *increase* in employment to 3,650 in 1982. Yet in Britain as a whole, through the normal process of capitalist competition and reductions in labour time, the result of the rise of the Japanese was a major net loss of jobs. Matters were further worsened by the preference of the Japanese, on the grounds of quality, for retaining their traditional suppliers, for this also triggered off job losses in the electronic components sector as technological complementarities worked in reverse.

Despite the fact that they were direct competitors and all partly or wholly Japanese owned, the three television companies among the new entrants had different effects both within the plants and in terms of their impact on local labour markets. The most extreme contrast was between one of the wholly owned Japanese subsidiaries (company A) and an Anglo-Japanese joint venture (company B), whose shares of the UK television market were 6 per cent and 16 per cent respectively in 1982.

The strategic aim of company A was *gradual* expansion to strengthen the plant's competitive position in the long term and – against the usual expectations of branch plants – to develop advanced technology products in its own right. Modest progress

had been made in these respects and, of all the Japanese colour television companies in Britain, A had progressed furthest in
exploiting development research on teletext and viewdata. These
were originally British innovations which A had incorporated into
its products and exported to Japan. Although its process technology
and social organization were not as advanced as in its home plants,
A had introduced automatic insertion machines for components from
the outset. This was despite the fact that volume did not appear to
justify its introduction: product quality seemed to be the main consideration.[3] Furthermore, the status of the plant has been upgraded
with the addition of a television tube facility which supplied in-house
needs for 30 per cent of its European sales for this key component.
Output had expanded annually since start up (with the exception of
1983) and, while this was expected to continue for both the UK and
EEC markets, employment (960 in 1983) would remain broadly stable
because of increased automation and productivity growth. Within its
local labour market this plant was perceived as a novelty, principally
because it had declared itself against any redundancies. So, in terms of
received notions, this plant, now in its tenth year of operation, could
in no way be subsumed under the 'insecure branch plant syndrome'.

Company B, the joint venture, was very much a marriage of
convenience for both partners. The Japanese company had earlier
been refused permission to set up a plant in the north-east of
England, having been blocked by a corporatist alliance between
indigenous firms, trade unions and the National Economic Development Office. Although this protectionism was short lived it was
sufficient to persuade the company to forego a greenfield site and
enter a joint venture, in a plant dating back to the fifties. The
British partner, whose record on product development, product
innovation, quality and marketing of televisions was very poor,
even though they were part of a large diversified engineering group
was desperate to remain in the television market. At the start of the
joint venture in 1979, the time-economies of set production were well
above the UK industry average of around 6.1 hours per set. Since
1979, output has doubled, labour time per set has been reduced from
nine hours to two in 1984, and employment in 1986 stood at 900,
having previously reached a peak of 2,450 in the 1973–4 boom.
Throughout the 'marriage' some 60 per cent of components were
still secured from Japan and attempts to produce some components
in-house were largely abandoned for reasons of cost and technical
problems. The sets were sold with either partner's badge on, there
being only cosmetic differences between the two parallel ranges of
models, but those with the Japanese badge were priced £50 higher

than the 'British' ones and still managed to outsell the latter easily – a striking illustration of the different reputations of the partners!

The management styles of the partners could hardly have been more different, the British firm being renowned for its short-term horizons and 'accounting syndrome' and in television production for poor quality, the Japanese firm being the opposite on all these counts. Not surprisingly, there was a high degree of inter-corporate conflict; indeed it represented a poor advertisement for Anglo-Japanese cooperation and a justification for the Japanese preference for wholly owned subsidiaries on greenfield sites. These conflicts threatened the security of the operation and aggravated industrial relations problems. The Japanese partner finally took over the whole venture in 1985, after protracted labour disputes in which shop stewards had written to the Tokyo headquarters asking for a takeover – a telling point against the view that external control is necessarily seen as a threat to security of employment!

Despite the success in attracting four Japanese consumer electronics firms, the regional development authorities were highly disappointed that none of them has introduced the new generation mass market product, the video recorder, into the region.[4] Their local multiplier effects have also been largely limited to their wage bills, for they have done little local – or even British – purchasing.[5] This preference for importing components from Japan has much to do with the poor quality standards offered by British component suppliers. The Japanese managing director of the third television plant visited 50 suppliers asking for zero defects instead of the usual British one defect per 1,000 standard, only to be met – at least initially – with disbelief.

Electronic components

The region's 'success' in this sector has been more mixed. Six electronic components plants were examined. The two cases reported here belong to quite different causal groups and represent specific responses to the dominant trends in the industry, particularly as regards technological and product innovation, corporate status, employment and maturity.

Company C is the only major employer in our survey that has its headquarters in South Wales, where it has been operating for 36 years. This company was originally a producer of electromechanical components, almost entirely reliant on the consumer electronics end market. Since the latter was a 'boom or bust' market (induced in part by 47 fiscal changes affecting credit terms in 25 years), the company consciously sought to diversify. A combination of foresight and necessity, in the shape of the recession after 1973 and the falling number of components per set in televisions, stimulated a bold diversification strategy. By 1981 its traditional consumer market accounted for only 7 per cent of turnover, while electronic subsystems for the MoD accounted for 12 per cent and professional and industrial equipment markets represented a further 23 per cent.

This diversification strategy was greatly facilitated by a licensing agreement with a US electronics company and by a series of acquisitions in targetted products and technology. (Two of these acquisitions have been along the English M4 corridor.) The major obstacles to this diversification have been associated with the recruitment of managers and R&D teams since incumbent personnel proved unable to make the transition to new technologies and product markets: recruitment problems were attributed in no small way to the received image of the region as one of 'coal, steel and militancy'.

Symptomatic of the hierarchical spatial divisions of labour discussed in chapter 8, the R&D work for C's most innovative product (a microprocessor-based control system for the automotive industry) was performed at its Berkshire facility where its advanced engineering centre is based. Despite its headquarters in South Wales, C has been obliged to adapt to the prevailing location patterns of top electronics skills.[6]

Interestingly, *product* diversification was accompanied by a *spatial* diversification so that its fastest-growing activities are not based in its three plants on the coalfield but, increasingly, in two plants in southeast Wales – within the WDA's golden triangle in Gwent (see figure 9.1) – and in its two plants in Berkshire. Spatial diversification has also been stimulated by its recent emphasis on small plants (in which the optimum size was considered to be 300 employees) and by its attempt to seek out non-coalfield labour markets with more flexible labour practices (see chapter 10). However, new employment growth in these less mature plants has not been nearly sufficient to offset the job losses associated with diversification and recession: total employment peaked at 3,550 employees in 1974 and fell to 1,960 in 1983 and, of all the sites, the headquarters plant bore the brunt of these losses. In other words, while company C has undoubtedly been very successful in corporate terms, this need not be wholly synonymous with regional health. And given that the company was one of the few to have plants in the middle of the coalfield where unemployment is worst, instead of on the fringes, the regional consequences were particularly severe.

Company D operates in a completely different context. It is of interest both as a unique venture in itself and as an illustration of the constraints and demands of the spatial division of labour in electronics. Established in 1978 in an attempt to provide Britain with an indigenous source of mass market standard chips – a market from which existing domestic firms had retreated to concentrate on customized products – this company hopes to catapult itself into advanced product markets. For this purpose D felt obliged to position most of its early R&D (especially development) facilities in the USA because this combined leading-edge semiconductor firms as well as a major pool of process engineers, of which there was a chronic international shortage. Equally important were the highly sophisticated end-user firms in the USA which, as we saw in part II, are reluctant to place contracts with foreign firms lacking a significant US presence. This locational behaviour speaks volumes about the perceived status of Britain as an environment in which to develop truly leading-edge innovations in this field; it also illustrates the fact that indigenous ownership does not eliminate technological dependence.

D has three major sites: R&D and production facilities in the USA; a British management and R&D centre in England; and its main volume wafer fabrication plant in South Wales. Most assembly work is subcontracted out to the Far East. In 1984 its employment in the USA stood at 750, with some 650 in the UK, though the American plant accounted for £34m of its £37m turnover. Unlike most plants in the region D has been able to buy its way out of the constraints imposed by its local labour markets. This, together with the attractions of working with advanced technology, has enabled D to recruit key personnel from all over the UK to its South Wales plant. Recalling the turbulent environment faced by any firm trying to enter mass production (see chapter 3), D's performance has, not surprisingly, been erratic. Although it has had some successes with its leading-edge, high-performance products, including a revolutionary computer-on-a-chip which permits parallel processing, its product range and depth have been too limited to give it any security. It has also been plagued by shortages of long-term investment funds, and although it has now been taken over by an established medium-sized firm, its difficulties in this sphere are far from over. In this respect D epitomizes the vulnerablility of British manufacturing, exposed by the notorious obsession with short-term profitability of British financial capital, in surviving in a volatile international competitive context.

Most of the other components firms specialized in subcontract work, for example, making printed circuit boards and assembling them with components. As such their prospects were bounded by those of larger firms and by the latters' decisions regarding the economics of in-house versus subcontracted production.

Microcomputers

The British home microcomputer industry appeared to be the most robust in Europe in the early eighties. Many commentators forgot what tends to happen to infant industries and took the dramatic growth of firms like Sinclair Research and Acorn during the early eighties as indicators of the possibilities of advanced-technology-based small firms as a 'vehicle for regeneration via innovation' in the older industrial regions (see Rothwell, 1982).[7] Certainly the growth of the market in Britain was spectacular: sales only began in earnest in 1980 and trade estimates suggest that it grew in value terms by between 40 and 50 per cent in both 1982 and 1983. In view of this it was sometimes compared hopefully with the youthful US semiconductor industry of the 1960s. However, the micro market is rapidly maturing and established computer firms – which initially treated home micros as 'toys', and thus beneath them – have belatedly entered the lower price end of the market.

Three firms in South Wales had some involvement with micros, two of which were subcontract assemblers for other firms (one of these being firm C), and the third, company E, which produced its own model. An offshoot of a former toy producer in Wales, E had its headquarters in the region and employed 270 in 1983. Its short life is typical of a small infant-industry firm. Its first model proved extremely successful with unit sales of 82,000 (65,000 in the UK) in its first 16 months. Nevertheless, its position was always highly precarious because of

1 fierce price-cutting,
2 an inability to attain volume production fast enough to compensate for declining (and already low) margins, and hence constant cash flow problems;
3 undercapitalization,
4 short product life cycles (maximum two years) which meant that new products had to be developed as soon as others were launched on the market and
5 a failure to construct or insert itself into a wider marketing system. Management felt that the future belonged to those (such as IBM) who were able to combine low-cost production with marketing strength.

Solving many of these problems would have undoubtedly involved reduced independence and greater external control by other firms. As it was, E was totally dependent on more powerful subcontractors who supplied the central processing unit, the cassettes, cartridges and cases. The company never became an integrated producer; bought-in materials accounted for over 80 per cent of the ex-factory cost of each micro. However, it was something more than an assembly-only operation because it combined this with an in-house development team (11 engineers backed up by 17 technicians) which was seen as its greatest asset. Nevertheless, the company's attempt to move into more powerful business micros and into software failed and the company collapsed in 1985.

All in all, E represents a salutary reminder of the problems confronting small firms in new technology ventures, especially where they start out in highly competitive mass markets, and a warning against the hopelessly idealized views of such firms current across much of the political spectrum.

Telecommunications

The main telecommunications plants in South Wales clearly illustrate the dichotomized structure of the industry. Two plants, F and G, were involved in pre-digital switching technology for protected public markets but were threatened by the transition to digital exchanges, while a third, H, a branch plant of a multinational, operated in a quite different causal context and had achieved rapid growth in PABX markets.

F and G were branches of diversified British companies which had operated in the region for over 40 years. They shared an identical corporate status as overspill plants and both concentrated on the more mature switching equipment within their companies (i.e. Strowger, Crossbar, TXE2 and, more recently, TXE4). Clearly, these plants constitute the very stuff from which the 'branch plant stereotype' is made, though as overspill factories they were in a minority.

Plants F and G had been under threat of closure for a decade as the market for pre-digital switching equipment contracted. Ironically, they owed their survival to repeated delays in the introduction of the fully electronic System X exchanges which prolonged the life of older technologies. But also important was the fact that these (largely female) workforces were among the most flexible and disciplined in their companies. (Some other plants in England making these semi-electronic exchanges have been closed.) Nevertheless F was to have been closed in 1981 but survived because it was transferred from the company's switching division to its transmission division, thus producing different products, and because a tutored workforce was preferable to expansion in the West Midlands where industrial relations were troublesome. Plant G, whose company is more dependent on semi-electronic exchanges, continues to concentrate on its traditional activities and, unless its function is redefined, as with F, its future is doomed. Not surprisingly, employment has halved in both plants since 1975 and their fate now hinges on how product innovation is managed.

The third member of this subsector (plant H) represents a stark contrast to F and G in terms of its technological status, corporate role and product market. This North American company established a facility in the UK (in 1981) for two main reasons: the UK was its largest market for PABXs in Europe, and the liberalization of telecommunications presented it with opportunities unique in Europe. A third factor was the quality of the UK's software engineers. South Wales was selected because it combined regional development grants with access to London and proximity to British Telecom, its major customer.

Unlike F and G, H in no sense conforms to the branch-plant stereotype: the South Wales facility is its headquarters for Europe, the Middle East and Africa, as well as housing its European R&D and marketing centres. In fact, H is unusual for a foreign-owned plant in so far as its R&D component really is *research* and development, and not just secondary or adaptive development work: unlikely as it may seem, South Wales is one of its three R&D centres and the only one outside North America. However, the contradiction with our claim about the character of industry in South Wales is more apparent than real for H is only just inside the Welsh border and hence enjoys some of the advantages of the M4 corridor and largely escapes the negative images of the region which deter professionals from working further west on the coalfield (see figure 9.1).

The unusual corporate and technical status of H was such that its occupational structure proved to be the least 'headless' of the Welsh plants: managers, scientists, technologists and technicians accounted for nearly 30 per cent of its total employment of 660. As a new plant it is expected to increase employment to around 800 but not beyond this ceiling. This is not just because of technical factors but also because smallness is now widely considered to be most conducive to manageable social relations.

The electrical products subsector

While virtually all the older plants in the region had shed labour during the seventies and early eighties, the process was particularly marked in firms making *electrical* – as opposed to electronic – goods, e.g. washing machines and light bulbs. However, shifts in product technologies were only a part of the reasons for their decline. Consider plant I for instance: this is a long-established foreign-owned branch plant making washing machines, whose employment had fallen from over 5,000 to 2,200. It had neglected product innovation – in particular, in lacking a machine with electronic controls – and had weaknesses in process technology and work organization. But these deficiencies only became apparent when exposed to a more pervasive problem: the massive degree of overcapacity in the European industry that built up in the seventies. In part, this was a result of the recession depressing demand, but it was also due to several major European firms seeing the potential of rising optimal economies of scale and increasing their capacity during this period. However, as is usually the case in crises of overproduction, competing producers were very unequally equipped to cope with overcapacity. Some Italian firms had taken the lead by setting up giant mechanized plants using cheap labour and their exports flooded into the UK market. Consequently plant I had been beset by a severe financial crisis since 1978. In the sixties and early seventies it had been too busy 'chasing volume' in an expanding market to worry about labour productivity and product innovation. Now, in response to its difficulties it was planning the introduction of computer-aided manufacturing (CAM) and trying to change its management–labour relations.[8] In short, the case of I shows both how the fate of plants and firms is interdependent, lying partly, though never wholly, within their own hands, and how the hidden hand of the market reflects the actions of other firms as well as shifts in demand, it being clear that I's problems were exacerbated by cost-cutting of the Italian producers.

While the influence of many of the processes identified in part II and chapter 8 was evident in most of the above cases, there was also a large minority of firms operating in other markets or doing subcontract work, e.g. a firm making video tape, another making parts for the automotive industry, a producer of torpedoes and sonar equipment, a firm making special batteries for burglar alarms, a printed circuit board firm doing subcontract work for computer firms, another subcontractor assembling boards for computers and microelectronic controls for cars and so on.

Stagnant or falling employment was dominant in these firms as a result of a variety of factors, most of which are familiar in the industry:

1 the effects of switching from electrical to electronic products (as in the automotive supplier, though with the decline of the British car industry depressed demand was important in this case too);

2 reduction of labour content with the development of microelectronics in general and the introduction of automatic component
insertion and test equipment (as in the subcontract assembly firm)
and robots (in the video tape plant);
3 intensification of labour, as we shall see in chapter 10.

Conclusions: development in or of the region?

The characteristics of the firms reflect both the general nature
of the industry (e.g. the contrast between consumer electronics
and telecommunications in terms of degree of internationalization),
the particular place of South Wales in the international and intra-
national divisions of labour, and, as we shall see in chapter 10,
the particular nature of the South Wales context.

Overall, then, the electrical engineering industry in South Wales
consists overwhelmingly of manufacturing, particularly mass and
batch production. Nearly all the plants are branch plants, but this
does not of itself signify either insecurity or absence of R&D as
the branch plant stereotype would again imply. On the first issue,
most plants do not perform 'overspill' functions, duplicating work
done elsewhere, but play unique roles within their parent firm
and hence are less dispensable than the branch plant stereotype
would imply. Secondly, it is not the complete absence of R&D
personnel but the predominance of routine occupations which
typifies the plants in South Wales; several plants do a limited
amount of development work, mostly related to incremental process
improvement. Although the plants are often the last to benefit from
technology transfer within their respective firms, they are still the
major source of innovation in the region.

In other respects, the region's industry largely conforms to the
stereotype. In particular, a large majority of the workers in the
South Wales firms are semi-skilled or unskilled. This lack of
front-end activities reflects both the region's social composition
and its consequent unattractiveness for professional and managerial
workers. Even though the numbers of such workers required were
small – itself a reflection of adaptation to local circumstances – virtually every firm complained of difficulties in recruiting such people.

In this regard it is significant that the two firms which differed
most from the regional norm in having large numbers of front-end
workers were located east of the coalfield, in the less-industrialized
'Gwent triangle', closest to the English M4 corridor. One of these
was firm H. More exceptional still was an R&D branch of a defence

firm which had been shifted out of Berkshire in order to escape labour poaching and housing problems. Its young workers could find affordable housing more easily in Gwent than in Berkshire (particularly in the new town of Cwmbran), though a minor 'problem' was the shortage of housing in the £70,000+ bracket for executives (1983 figure) – the converse of the housing situation in Berkshire! (See chapter 12.) With its young, male, besuited workers and its office-like atmosphere this plant was very much in the (English) M4 corridor mould and quite unlike the other Welsh plants,[9]

Given the social composition of the region it is not surprising that its main prospects for growth lie with inward investment rather than new firm formation. Yet new plants starting up in the region were only creating new jobs at a rate sufficient roughly to balance the losses in the older plants. However, as was so patently the case in consumer electronics, the leading entrants have also been partly responsible for exposing the weakness of the established firms and, indirectly, bringing about job losses in them.

The combination of output growth with stable or declining levels of employment does not appear to be a temporary phenomenon caused by recession or spare capacity. When asked what the employment effects of 20 per cent output growth would be, most managers felt that it would be between 0 and 5 per cent. Significantly, the only major exception to this trend – apart from very recent entrants still building up – was the defence R&D plant mentioned earlier, whose project work allowed no scope for economies of scale.

The diffusion of growth through local purchasing and subcontracting is inhibited by the tendency of large inward investors to source components in-house (i.e. usually from overseas), by the restricted number and range of local firms (itself an effect of the region's social composition and industrial history) and by the lack of quality assurances which they, along with many British firms, offer. Moreover, given the overwhelming dominance of back-end hardware production, the industry offers few of the opportunities for local software houses that exist in the M4 corridor, though their absence also has much to do with the evident distaste of software specialists for living in old industrial areas.

In other words, the local multiplier effects of the industry must largely be restricted to the effects of the spending of wages though even here these are limited. But this is not all there is to the assessment of the industry's impact on the region for, given widespread incidence of jobless growth in most of the older plants (i.e. those over 10 years old) and the precarious nature of new firms like D and E, the actual performance and prospects of the firms are absolutely crucial for the region: hence our concern to look into the determinants of their performance in some detail. Indeed, in retrospect, it seems utterly extraordinary that the older (and even some recent) literature in regional studies could have ignored this and treated plants largely as black boxes to be enumerated and correlated with regional aid or whatever, though no doubt this neglect was not so obvious during the post-war boom when unemployment was low and plant closures few.

The dependence of developments at the regional level on international competitive and production systems could hardly be clearer. Half of the plants were foreign owned and all but a few of the oldest of these were leaders in their fields. As publicists such as the WDA have tirelessly reiterated, the array of firms in Wales contains some of the most prestigious names in the electronics industry worldwide. Yet for analytical, as opposed to propaganda, purposes it is wholly misleading merely to register this international aura and then proceed to facile conclusions about alleged similarities to the regions of origin of the inward investors. Contrary to Welsh Office claims that Wales is comparable to Boston's 128 corridor in the USA, the Welsh electronics industry most certainly occupies a different position within the international and corporate divisions of labour to that of Boston (Morgan, 1987b). In short, we must reluctantly conclude that this is more like development *in*, rather than development *of*, the region of South Wales.

10

Social Innovations 1
Electrical Engineering in South Wales

As we saw in chapter 9, the new entrants in South Wales include some of the most prestigious names in the electronics industry. In relation to Wales and Britain many of them are innovative not only in their technology but in their social organization. Yet, on the face of it, the regional legacy of South Wales is hardly inviting to such firms. In order to see whether this is the case, we must look at the forms of social organization of work at the point of contact between the old and the new spatial division of labour. How can leading modern companies operate in such a context and what difference do they make to traditional local work culture? However, there are other aspects to the context which suggest that the social innovations may have wider significance, namely the roles of accelerated technical change in restructuring and of recession and mass unemployment in conditioning workers' behaviour.

In this chapter we examine how the firms and their workers have adapted to one another and assess the significance of the resulting management–labour relations, both for company performance and for labour generally. We begin with an outline of the characteristics of the workforce in electrical engineering in South Wales, comparing it with that of the traditional industries. We then detail the reconstitution of management–labour relations in the industry and the response of trade unions in this most labourist of British regions, concluding with a brief analysis of the spatial recomposition of the industry in South Wales.

The workforce: skills, gender and pay

Given its largely 'branch-plant' character, South Wales is not in the running as a location for top calibre skills. Partly because of the strength of unionization, most firms there are unwilling to pay a premium to attract them for fear of triggering off wider disputes over pay differentials. However, the status of skill as a location factor is too often equated with its technical sense, based on formal qualifications etc. Yet there is another conception of skill that is now increasingly employed within the industry. This is rarely formally defined, but it seems to refer not so much to the technical qualifications of employees but to their qualities as 'good company employees' in terms of attendance, flexibility, responsibility, discipline, identification with the company and, crucially, work-rate and quality. This conception refers, then, to the behavioural qualities of labour. The use of this concept of skill is most common in the Japanese plants: in one of these, some new uninitiated Japanese managers misconstrued the traditional British categories of skilled, unskilled and semi-skilled as meaning good, bad and indifferent! On this concept, 'shortage of skills' often means the lack of workers who have fully adjusted to corporate standards. As we shall see, there are important spatial variations in the incidence of such behavioural skills within the region, at least in the eyes of management.

One of the most striking contrasts between the electrical engineering workforce in South Wales and the traditional sectors concerns the gender composition of the workforce: women form 45 per cent of the workforce in the former (58 per cent in electronics) but only 5.6 per cent in mining and metal manufacture (1981 ER II data). Apart from the clothing and food industries, electrical engineering has the highest level of 'feminization' in South Wales manufacturing. The fact remains, however, that the legacy of masculinity associated with coal and steel lingers on and hence attitudes to women's employment are less liberal than would be expected in labour markets more dominated by employers of women. There seemed a large measure of consensus among managers, workers and even union officials as to what constitutes men's and women's jobs: the tags 'girls', 'women', 'lads' and 'men' being applied selectively to the holders of particular jobs as if it were second nature, although when questioned explicitly on gender, managers and union officials would of course generally deny discrimination.

The large proportion of women workers reflects the conjunction of the dominant branch plant status of electrical engineering in South Wales and the overwhelming restriction of women in operator and clerical grades. While in Britain as a whole the influx of women into the industry in the sixties and early seventies has been reversed by contraction, automation and the reduction of manual assembly work, the latter trend has been offset in South Wales by the influx of new branch plants.

A further reason for the declining attractions of female labour was equal pay legislation. Even though employers have minimized (sometimes with union complicity) the full potential effects of such legislation – by redefining work content and through job evaluation schemes, which often institutionalized gender segregation in ways that did not inflame parity claims – some companies in South Wales felt that equal pay had a penalizing effect much greater than that associated with the withdrawal of the regional employment premium.

It must also be said that the female stereotype is more powerful where it is ratified by female workers themselves, rather than where it is imposed only by managers and unions. For instance, even allowing for informal slanting of job vacancy advertising there is a considerable degree of segregation at the application stage, which suggests that individuals *self-select* according to prevailing concepts of gender when applying for jobs. In such cases, the community actively reproduces established models of gender which companies find congenial. We were also struck by the fact that this self-selection is not confined to work which is familiar to the community (e.g. 'women's jobs' in wiring and soldering) but extends to types of work with no precedent (e.g. semiconductor plant operatives). This societal reinforcement of gender stereotypes in the workplace is sometimes underemphasized in radical accounts so that it appears that company discriminatory preferences encounter more resistance than is actually the case: much of the 'dividing and ruling' happens within the community.

However, sexism does not always work in the corporate interest. Some managers recognized that the solidarity of male craftworkers such as fitters and their sexist attitudes towards female operators (whose bonus payments often depend on their cooperation) were an impediment to high productivity, yet the firms lacked the will or means to desegregate the occupations. It is possible that although divide-and-rule policies always carry the risk of creating frictions which lower productivity, desegregation might create more problems from capital's point of view.

The importance of gender is further illustrated by the fact that two of the biggest industrial disputes in the industry in Wales centred on equal pay and sex discrimination. In 1976 a protracted strike for equal pay at the GEC Telecommunications plant in Treforest was led by women. And, in 1981, women at Hoover prosecuted a successful campaign against both management *and* 'their' union who together sought to deprive them of seniority status and access to higher skilled jobs. Again, in 1983, women at GEC-Hitachi initiated a walk-out against management attempts to reduce bonus payments. This may not be simply because operators tend to be harder hit by management offensives: as some regional union officials attested, although women are more reluctant to take action, once they do they are noticeably more combative and solidaristic than men.

Generally, we found little evidence that union attitudes towards women workers were changing in more than superficial terms. One male official we interviewed was frankly misogynistic, and others often reproduced and cemented the 'female stereotype'. Interestingly, some also said that awareness of 'women's issues' was greater among trade unionists in London and the south-east than in Wales. Clearly, from the point of view of women, the assumption that South Wales is a 'radical region' carries some ironic twists. The reality is complex: traditional masculine culture, trade union male strength and low female activity rates in one sense act against women's interests, but the pro-union culture and the consequent relatively high rates of unionization amongst women noted earlier provide a possible base from which they can fight back.

Published information on earnings in electrical engineering is either coarsely aggregated or extremely patchy at a detailed level. However, it is clear that basic rates of pay for semi-skilled workers (usually women) are about half those of miners and steelworkers. For example, in 1983 basic rates of pay for semi-skilled workers were between £76 and £88 per week, though bonuses are paid in some cases. At the same time the median weekly earnings for miners in Britain were £183.2 per week and, for male manual workers in iron and steel, £165.2 per week.[1] This means that while such jobs in the new industries may be giving some women a chance to earn a wage for the first time, their rates of pay are low, not only relative to earnings in the traditional industries but in absolute terms, though they are certainly not unusual in Britain. Households wholly reliant on a single income of this level fall below the official poverty line.[2]

The reconstitution of management–labour practices

Three contextual tendencies are vital to the explanation of management–labour relations in the eighties:

1 the deep recession between 1979 and 1981 and the emergence of mass unemployment;

2 the acceleration of technical change;
3 the increased emphasis upon inward investment as a vehicle for reindustrialization in the UK generally and in its peripheral regions in particular.

Together, these tendencies have encouraged a reassertion of the managerial prerogative and the latter has been enormously facilitated by the advent of the Thatcher Government in 1979. The managerial offensive in the UK is clearly uneven within and between industries but, thus far, its most dramatic instances have occurred in the public sector industries of coal, steel and motor vehicles (Morgan, 1983a). While it assumes many forms, it is principally designed to reverse 'overmanning' and 'restrictive working practices', to impose stronger disciplinary norms and, where possible, to enlist consent and identification, preferably of the *active* type, because – as Gramsci well appreciated – this is not an unimportant condition for the reproduction of any hegemonic system. As we shall see later, the mobilization of consent is fast becoming a key element of the new 'management paradigm', itself a backlash against conventional management practice in the USA and UK especially (Littler and Salaman, 1984).

Management and labour practices are interdependent but together they vary significantly between industries, between firms in the same industry, between plants of the same firm and even within plants between different categories of workers. Notwithstanding these variations, the following characteristics have been widely observed across large sections of British manufacturing industry (see Pavitt, 1980; Brown, 1983; Williams and Williams, 1983):

1 management assumes little interest in the technical and social details of shop-floor production;
2 management and labour are relatively conservative as regards innovation and the former operates with short time-horizons with respect to investment, R&D and profits;
3 management communicates with the workforce via a 'frontline' of shop stewards (buoyant labour and product markets obscured the fact that shop stewards' facilities lay largely in the hands of the employer, while the relative absence of written agreements on working practices made it easier for managements to reverse 'custom and practice' in the absence of workplace opposition);
4 unsophisticated recruitment methods;
5 multiple and complicated pay structures and, related to these, rigid job demarcation;
6 low productivity and low priority accorded to design and quality;
7 several unions per plant;

8 little effort to encourage worker identification with firm or plant;
9 disregard of workers' knowledge of the labour process in problem-solving;
10 very hierarchical structure within management with strong status inhibitions characterizing management–labour interactions.

These characteristics, relevant mainly to large unionized manufacturing plants, have contributed to the relative weakness of British management and the strength of shopfloor organization, especially among male craft workers. Many of these features were evident in the older British and American plants, and our impressions were confirmed by shop stewards and union officials.

The main social innovations associated with the more recent entrants (post-1973) are as follows.

Recruitment procedures

Recruitment procedures are far more rigorous and tend to be carefully tailored to the needs of particular firms. This is even the case for unskilled and semi-skilled workers: as we shall see, this makes sense for capital, for it is hardly rational to recruit direct production workers – on whom productivity very largely depends – in a casual manner. Interviews tend to be carefully planned and thorough rather than perfunctory as was traditionally the case in British firms before the recession. Tests for dexterity and character are common and references are carefully scrutinized. Information is often sought on family background and hobbies, to use as indicators of possible attendance records and behavioural skills. One firm refused to recruit single mothers, preferring workers with stable two-parent families without pre-school age children. Others relied more heavily on tests and probationary periods in their selection procedures, but in both cases the aim was the same: to select workers who had or would develop the necessary behavioural skills and who would have minimum distractions from the domestic sphere.

More generally, the increasing importance attached to recruitment has been both allowed and necessitated by the recession. Some managers in older firms noted the contrast between the present and the heyday of the sixties and early seventies when demand was high (and less affected by foreign competition) and when, at the extreme, they were 'chasing volume'. Workers would be hired (and fired) with little care, for there was relatively little pressure on them to worry about productivity or quality: they could virtually sell all they could make. The relative buoyancy of the labour market conversely meant that although they had less scope for choice, high turnover of labour enabled them to compensate for 'excessive hiring'.

After 1974, and especially after 1979, these conditions were reversed: product demand tended to be more stagnant and competition, in terms of quality at least as much as price, intensified considerably. Labour turnover fell and firms found themselves saddled with labour which was not only surplus to requirements but low-skilled in the behavioural sense (see Income Data Services, 1984). As one manager in an old firm put it, one of the main lessons that the recession taught British management was that it had for too long regarded labour as a 'cheap commodity'; in the short run, it *is* cheap and easy to buy, but treating it as such encourages complacency and lack of innovation. Extra recruitment is no longer a reflex response to increasing output. The major recent foreign entrants, with their longer-term planning and profit horizons, tended not to hire labour reactively as markets fluctuated, as did the older British and American firms, but in accordance with long-term corporate strategy.

The result of this shift into recession and of the influx of foreign firms is to make employers more reluctant to employ workers from established firms and particularly from the coal and steel industry, except where they are obliged to recruit the latter as a condition of obtaining European Coal and Steel Community Grants: such workers are otherwise doubly disqualified because of inappropriate technical and behavioural skills.

Flexible work practices

The most distinctive feature of the work practices of the newer firms is their emphasis on flexibility. Reduced demarcation, allowing wider margins of discretion over job allocation, was apparent not only in comparison with the traditional coal and steel industries but also relative to the longer-established electrical engineering plants. Demarcation – effectively a way in which unions inflated employment in the long boom – was now under attack from recession, new foreign firms and new management philosophies. Such flexibility was most pronounced in the wholly Japanese-owned plants in the consumer electronics sector. Although their productivity levels were generally still below those of plants in Japan, they had an advantage over indigenous producers not only through superior quality (and hence less time wasted on rejects and correcting faults) and through higher work speeds on individual tasks but through more rapid transfer of workers between tasks.

This flexibility was facilitated not only by the recruitment policies already noted but by a refusal to countenance multi-unionism and by institutionalizing flexibility in written agreements: hence the GMBATU[3] in South Wales says, in its 'model agreement', that a condition of employment is that each employee accepts 'the direction of the company's management to perform *any* kind or type of work within the employee's known abilities' (General, Municipal, Boilermakers and Allied Trades Union, 1981). This trend toward the 'flexible worker' is in part responsible for the loss of job descriptions, one implication of which is that technical and behavioural skills are becoming more firm-specific and less transferable between firms (Brown, 1983).

A parallel shift towards greater flexibility is also emerging in the higher technically skilled occupations in the shape of the multiskilled engineer or technician, although in this case technical convergence between different engineering skills in the industry (e.g. electromechanical, electronic, software) is a contributory factor. (However, while many firms seek such workers, few in Wales devote many resources to training them, yet still complain of a shortage.) According to Takamiya (1981), some of the Japanese plants insist on job rotation among managers so that they work in different departments and temporarily do production work, the aim being to restrain departmental empire-building and facilitate both horizontal and vertical communication.

We shall later argue that such developments, when coupled with others detailed below, signify a partial rejection by 'progressive' management of Taylorist wisdom − i.e. the division of labour into highly specialized, simplified and deskilled tasks − because it is associated *inter alia* with internal labour market rigidities and, therefore, with diseconomies (see Abernathy et al., 1984). However, as we shall see, a policy of improved flexibility cannot hope to be successful if it is pursued as an isolated objective; it requires a *package* of related policies including rigorous recruitment procedures, single union status, simplification of pay structures, squeezing of hierarchies and new management practices.

Simplified pay structures

Just as 'single status' is a device used to *induce* flexible work practices, simplified pay structures (rather than higher pay) help to *sustain* them. In larger established plants, the multiplicity of payment structures, together with uneven access to bonus opportunities, produced marked inter-job rigidities so that workers were extremely reluctant to relinquish their 'patch'. In some of the new entrants there was no bonus system at all because this placed a premium on speed over quality. Where bonus systems were employed, they were monitored very closely not least to ensure that all operatives had equal access.

Piecework and other bonus systems have long been accepted as crucial motivational devices but, in the more sophisticated repertoires, these were seen as wholly inadequate. What seems to be emerging in some of the 'pioneering' plants is an attempt to uncouple an individual's pay from the going rate for the job (and, in a minority of cases, negotiating with each worker individually!). This strategy, rendered more feasible in small non-union environments, means that rewards are not simply determined by job content or technical skill but, also, by the behavioural skill of an individual. This strategy was not confined to non-union plants. The firms which pioneered single-union status in South Wales felt that while it was indeed problematical it was nevertheless essential to legitimize the practice whereby employees having the same degree of technical skills could command different rates of pay. In effect, this meant attempting to win acceptance for the practice of moving workers for short periods onto jobs which normally commanded higher rates of pay but without paying them those rates.

Managerial control

Another distinctive feature of the new entrants is the extent to which they have circumvented the union(s) as a medium through which information is imparted to, and received from, the workforce. Among the more mature plants, management tended to communicate with the workforce indirectly via the shop stewards who, in effect, became privileged carriers of information. Management in the new entrants – many of whom believed that line supervision had been abdicated by UK managers, especially in the sixties – accorded a strategic priority to direct and regular forms of communication, using them for feedback and as a 'listening system'. One effect of this is to marginalize the role and status of the union(s) in the everyday life of employees: even routine grievances, traditionally taken up by shop stewards in Britain, are processed through immediate supervisors (see Income Data Services, 1984).

It is no coincidence that, in the most extreme of the traditional stereotypes we encountered, management was highly centralized and far removed from the details of the shopfloor, and consequently, the union was indeed a privileged carrier of information. Significantly, during its acute crisis between 1979 and 1982 (induced to a great extent by its neglect of product and process innovation, ignorance of direct labour productivity and recession), its major response – besides mass redundancies – was to decentralize its management structure so that it became more attuned to the direct production process. One of its major problems during this transition was to fashion a more *direct* relationship with its workforce and here it encountered great difficulty in circumventing its shop steward system. But the brunt of such changes does not fall purely on the workers; it involves managers reversing their tendency to see distance from the shopfloor as an index of status and it requires them to take over more of the responsibilities formerly ceded to foremen, supervisors and shop stewards.

While the older plants were beginning to appreciate the importance of behavioural skills, the new entrants recognized that these qualities were not to be 'discovered' but, rather, that they were the *result* of involved and knowledgeable management activity. While investment in advanced equipment (e.g. automatic assembly and testing) was a necessary condition for quality control, it was generally conceded that this was far from sufficient because of the imperative of 'building quality into products' at each and every stage of the labour process, rather than attempting to 'inspect it in' at the final stage. To this end, many of the new entrants had developed 'quality circles' or 'involvement teams', the basic aim of which was to enhance the quality of output by utilizing the knowledge and skill of the workforce. The effectiveness of quality circles varied but in all cases efficiency was said to depend on management communicating regular information, defining tasks, demonstrating the interdependence of tasks and training the appropriate team leaders.

Significantly, in many of the more mature plants it was felt that quality circles could simply be 'set up' as though the development and utilization of behavioural skills was a question of (employee) willpower and (management) exhortation. The Japanese in particular ridiculed this notion, arguing that quality circles etc. presupposed management calibre and a sufficiently tutored and trained workforce; although the Japanese plants used to receive enquiries from other firms on how they ran their quality circles, three out of the four Japanese plants visited during the research considered that they were not yet ready for them.

Further related parts of the package of new work practices have been the 'squeezing' of the vertical division of labour within plants and the harmonization and improvement of working conditions. The aim has been both to reduce the indirect labour force, and hence raise output per worker, and to allow more direct access of management to the shopfloor. This in turn presupposes greater flexibility, simplified pay structures and improved training in order to produce the behavioural skills – particularly self-discipline and responsibility for some decisions – necessary for making the jobs of the displaced grades redundant.[4] Significantly, one firm which tried to do this without a greenfield site and new labour had to abandon the attempt in the face of problems with middle management and supervisory skills.

The squeezing of the hierarchy is also intended to raise morale and identification with the company by reducing status differences. Improvements in working conditions and moves towards their harmonization across grades are also directed towards this end. Typical measures in the new plants include standardization of payment methods to salaries, sick pay and holiday entitlement, shared canteens, medical check-ups and membership of private health schemes. While unions might welcome many of these (even private health care in the case of the EETPU[5]) and find them congruent with their search for a move up-market and a new professional image as reskilling occurs in the industry, such innovations are also intended to marginalize unions by making them appear either unnecessary if they seek to guarantee what employers are already offering or anachronistic if they do not. As one might expect, the unions representing the lower-status workers were keener on these innovations than the higher-status unions, who had previously shown little inclination to liaise with the manuals – an example of divisions being reduced by capital rather than labour!

Naturally, such benefits have nothing to do with altruism and everything to do with sophisticated economic calculation. For example, one firm provided a taxi service to take workers to and from the dentist should they be unable to get an appointment outside working hours, but this is not merely convenient for the workers: it ensures that absences are minimized and taken for valid reasons. Nor do we want to give the impression that life for workers in the new entrants is easier than that for those in the established plants. On the contrary, the intensity of work is usually notably higher and managerial expectations of workers higher too. In the old television plant taken over by a Japanese firm line speeds had actually doubled. To facilitate this they asked every worker over 35 to resign, allowing them to nominate a younger relative for possible recruitment! The new workers, mostly teenagers, proved more able to keep up with the line and spoke dismissively of their elders' work rates.

Core–periphery divisions in the workforce

Divisions between primary and secondary workers are a long-established feature of capitalist firms but it is evident from our

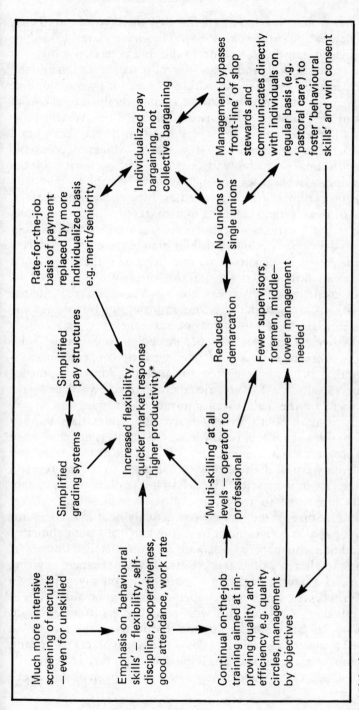

Much more intensive screening of recruits — even for unskilled

Emphasis on 'behavioural skills' — flexibility, self-discipline, cooperativeness, good attendance, work rate

Continual on-the-job training aimed at improving quality and efficiency e.g. quality circles, management by objectives

Simplified grading systems

Simplified pay structures

Rate-for-the-job basis of payment replaced by more individualized basis e.g. merit/seniority

Increased flexibility, quicker market response higher productivity*

'Multi-skilling' at all levels — operator to professional

Individualized pay bargaining, not collective bargaining

Reduced demarcation

Fewer supervisors, foremen, middle—lower management needed

No unions or single unions

Management bypasses 'front-line' of shop stewards and communicates directly with individuals on regular basis (e.g. 'pastoral care') to foster 'behavioural skills' and win consent

* If the firm achieves these, and provided all competitors do not do the same, thereby cancelling out the advantages, it should become more profitable and be better able to 'buy consent' of the workforce and make unions seem unnecessary.

Figure 10.1   The new management—labour relations

research and from other recent studies that the divisions are being made in new ways (Atkinson, 1984). The older plants tended to base the division largely on technical skills and perceived status: the higher the technical skills and status, the more secure the job and the more likely is an internal labour market to exist for promotion. But, strictly speaking, such a demarcation is not wholly rational from the point of view of capital because it disregards behavioural skills and allows non-economic criteria of status (which may have more to do with distance from manual work) to influence economic decisions affecting profitability. Using such criteria, many of the direct producers whose performance is most crucial to company success (e.g. operators) are categorized as dispensable. Significantly, a number of new entrants adhered to a conception of core workers which embraced a variety of occupations and ascribed strategic significance to developed behavioural skills among the direct producers and this in no way penalized technically semi-skilled operators. One American firm coined the phrase 'rings of defence' for its strategy of insulating core workers from the vagaries of the external labour market, 'surrounding' the core with rings of 'supplementals' on fixed-term contracts and increasing subcontracting as a further buffer. Divisive though this may be, the policy was openly 'sold' to the core workers as a way of winning their cooperation. Such policies might be seen as a step towards the Japanese practice of offering 'jobs for life' for a restricted group of core workers.[6]

These then, are the main social innovations emerging in the new electronics firms in South Wales; figure 10.1 summarizes the key interrelationships. They are, however, just tendencies, not completed projects; progress has been uneven, even within the newer plants, and we have been wary of the tendency of managers to try to present a progressive face in interviews. The delays in the older plants cannot be wholly attributed to technological backwardness and worker resistance, for some of the innovations actually hold attractions for workers. It has also been difficult to recruit managers not habituated to the traditional management–labour relations detailed above.[7] A further obstacle lies in the fact that the anti-Taylorist element of giving more workers a limited degree of responsible autonomy requires an increase in worker motivation, cooperation and goodwill. While it may be possible to develop this with new workers in a new plant building up to full capacity, it can hardly be expected in an established firm which is slimming down, making workers redundant, speeding up work rates and attacking union power! However, as we shall see, the role of trade unions has been far from obstructive.

Unionization

The labour traditions of South Wales were historically formed around the coal and steel unions. Although the practices associated with the traditional industries could not be identically reproduced in the electrical engineering sector, South Wales was still perceived by managers as 'one of the most heavily unionized regions in Western Europe', where belonging to a union is 'as natural as breathing'. Unionization remains pervasive in electrical engineering in South Wales and, although this may legitimately be read as an index of the persistence of labourist culture, little more can be read from unionization *per se* (e.g. 'militancy', 'activism'). Such a situation hardly fits with the ideology of high tech, with its close coupling with Thatcherist anti-unionism; for, in the glamorous world of new technology, trade unionism is seen as an anachronism redolent of smoke-stack industry and a barrier to progress. Nevertheless, all but two of the most advanced electronics plants in South Wales in our survey accepted unions, albeit in new forms. In the case of one major US multinational, the Welsh plant was its only unionized site worldwide, while several other firms admitted that they were breaking with corporate precedent in allowing British-style unions. More puzzling is the contrast with central Scotland where many leading companies have resisted trade unions despite the fact that much of the area has a similar industrial history to South Wales and, in 'Red Clydeside', a militant heritage.

A further angle on the problem emerges from a survey of US firms in Wales, in which managers offered very different readings of unionization (Maynard & Co., 1978):

1 strongly unionized but well disciplined by their union (58 per cent);
2 not strongly unionized but cooperative in settling disputes (34 per cent);
3 aggressive and not fully under union discipline (8 per cent).

Although the lack of a 'cooperative and un-unionized' category in this survey might be criticized it does puncture the widespread belief that unionization *per se* presents an obstacle to capital: whether it does depends on its *form*. (Some managers of branches of firms with unionized plants elsewhere in Britain said that they had found a more cooperative – and in one case a 'less sophisticated' – type of trade unionism than in London or the West Midlands.) Obviously, we must acknowledge that we have no way of knowing how many possible entrants to the region were deterred from locating there by unions and their reputation, but there is abundant evidence that unions have helped more than hindered those firms which have settled by providing a means to an 'ordered environment'.

Apart from the circumvention of trade unions referred to earlier, the most dramatic development in South Wales concerns the emergence of single unionism and, in a minority of these cases, this is accompanied by no-strike agreements. Although single-union status was pioneered in South Wales by Sony and the AUEW[8] as early as 1973, it is in the context of the post-1979 climate that the number and ideological significance of such agreements has increased. The major advantage, from a management perspective, of single-union status is that it promises more flexible work practices because demarcation lines (especially at craft and technician levels) are often institutionalized in separate unions.[9] Thus far there are at least eight single-union plants in electrical engineering, six of them Japanese. Their number will undoubtedly increase, less through *in situ* renegotiation than from new entrants.

Single-unionism in British manufacturing is most conspicuous in the (foreign) electronics sector and the Wales Trades Union Congress (TUC), anxious to accommodate company preferences so as to command a *spatial* advantage, has assumed a directive role in 'delivering' single-union status: while all unions may legitimately seek to recruit, once agreement is reached between a particular union and a company all others are obliged to withdraw. However, competition between unions of multi-union plants to deliver single-union status has been evident. While such behaviour seems profoundly irrational from labour's point of view, once one union starts to be successful in such a strategy, others are driven to do the same to counter it.

By far the most controversial innovation as regards management-union relations revolves around the so-called no-strike agreements, pioneered by Toshiba and the right-wing EETPU at Plymouth in 1981. The main ingredients of such agreements are normally single-union recognition, equal conditions for manual and office staff, flexible work practices, an advisory board of elected staff representatives (who are not necessarily union members) empowered to discuss any issue, negotiating procedures resulting in compulsory 'pendulum' arbitration (so called because the arbitrator must decide in one side's favour and cannot 'split the difference') designed to encourage moderate claims and union resolve to settle disputes 'without any form of industrial action'. One semiconductor firm initially declared itself against any form of unionization but relented after it won strategic concessions on the following: no strikes and no actions which interrupt the continuity of production; unionization to apply only to operators, clericals and junior technicians (while the advisory board embraces *all* workers); and union ratification of the company's prerogative to 'remunerate its employees on the basis of *individual* performance' (i.e. behavioural skills).

All too often, the management–labour practices associated with new entrants are simply reduced to single-unionism or no-strike deals when, for management, these are merely the most obtrusive insignia of a whole series of social innovations. The no-strike agreement is a brittle concord to sustain even for the most sophisticated managerial repertoires, but without the in-plant innovations noted earlier, together with the solvent of growth upon which much depends, no-strike agreements have little chance of being established, let alone sustained.

As might be expected, the no-strike issue has induced bitter internecine conflicts both between unions (each intent on establishing a presence in high tech, but few willing *overtly* to endorse EETPU-type concessions) and within the EETPU itself. For example, shop stewards feel a conflict of interest between their role as stewards and their role as advisory board representatives; they also feel that they are used as agents of the union bureaucracy, rather than as representatives of their members. Not surprisingly, EETPU officials have encountered opposition from their own shop stewards to no-strike agreements. Generally, we suspect that national officials are more in favour of single-union status than shopfloor representatives, as would be expected given its top-down corporatist origins. Indeed, most union bureaucracies appear to be adopting more corporatist forms of recruitment and liaison with management and marginalizing union organization at plant level. For example, in one Japanese plant with single-union status, the regional officials of the union have guaranteed that, in the event of any serious problem, plant management should communicate directly with the regional office where its problem would receive priority treatment, because (in the union's view) nothing should 'interfere with continuous production'. (Note that this is a plant *without* a formal no-strike agreement, which suggests that the controversy associated with these agreements lies partly in their *formally* acknowledged character.)

This top-down approach is most clearly identified with the EETPU and it is significant that a growing body of management favours this union above others: in cases where multi-union plants have become single-union sites it is invariably the EETPU which is retained. The reasons most often cited for this preference are its positive attitude to profits and flexibility, its use of its own resources to train and retrain for multiskilled grades and the fact that, because EETPU officials are appointed rather than elected, they can afford to take a firmer line with their members, thereby delivering a more disciplined workforce.[10] This is a truly ironic 'asset' given the Conservative Government's measures to force union officials into a more accountable position vis-à-vis their members.

New entrants – particularly from overseas – possess a distinct advantage over established firms in realizing the above innovations because labour lacks the bargaining power of existing involvement and precedent. Whereas a new union agreement in a multiplant firm like GEC might challenge precedents in plants across the country, a new entrant can start with a *relatively* clean sheet. We say 'relatively' because local characteristics still matter; none of the Japanese plants, for example, has more than a very limited Japanese character. In delivering social innovations to the new entrants, the unions acknowledge that they have conferred competitive advantages on such plants. Significantly, the EETPU is only too willing to replicate its innovations with British firms but, it claims, British management tends to be too conservative, too secretive and not sufficiently interested in consultative procedures (*The Engineer*, 1984).

In South Wales some of the innovations (e.g. flexibility, single staff status and relative job security) have been embraced with alacrity, especially by union officials. For local communities, workforces and union organizations, overseas inward investment is perceived not in terms of a burgeoning problem of external control but as a welcome invasion. While this may seem a rather unexceptional response in a region afflicted by long-term unemployment, it often tends to be ignored so that external control is simply read as an index of the regional problem (see, for example, Brown (1975)). Undoubtedly the individualistic or sectionalist desires of specific groups of workers to be associated with firms which have a track record of exceptional growth and market leadership encourages such favourable agreements with their employers, regardless of possible unfavourable consequences for the wider labour movement.

Finally, notwithstanding the lack of resistance to the innovations from trade unions, it still remains difficult to explain why the experience of South Wales should differ so strongly from that of central Scotland where, despite a similar regional legacy, the electronics industry is predominantly un-unionized.[11] Two possible explanations in particular merit consideration. First, the Scottish new towns have been privileged locations for inward investment in the electronics industry, particularly East Kilbride, Livingston and Glenrothes: it is of some interest that these new towns consciously recruited the most 'respectable' families and so never became centres of overspill from Glasgow, as was the case with Cumbernauld.[12]

Second, the Scottish electronics industry is far more dependent on US entrants, and IBM, Motorola and National Semiconductor have long been prominent non-unionized 'flagships' whereas, in South Wales, the most conspicuous new entrants – the Japanese plants – have all respected unionization, albeit in distinctive ways. IBM's role in Scotland seems to have been especially important. According to Foy (1974), the company is reputedly hysterically anti-union and no part of the company has been unionized; the few workers who tried to organize plants in the USA have been fired. Even where, as in Sweden, many workers join unions, IBM does not bargain directly with their unions, though it respects industry-wide agreements. Yet IBM has never had a strike. The avoidance of strikes and unionization is par-ticularly remarkable in IBM's Greenock plant – an area traditionally known as Red Clydeside – while ICL and two other major American electronics firms which entered Scotland in the 1950s – Honeywell and Burroughs – have had both.[13] In 1973, ASTMS's Clive Jenkins launched a campaign to unionize these firms and quickly gained 1,100 members, yet he only managed to recruit 10 of IBM's 2,000 workers at Greenock. According to Foy, unions are kept out by good pay, a well-publicized and effective grievance procedure and the exceptional job security. That this is not merely public relations is suggested by the fact that labour turnover in the company is less that 3 per cent per year (Bassett, 1986). However, while Foy cites non-autocratic management as another factor, she notes that any individual who does join a union is blacklisted and debarred from promotion. Nevertheless, such has been the success of IBM's non-union pol-icy that some observers argue that it has significantly weakened trade unionism in west central Scotland and encouraged other anti-union firms to set up in the area (Dickson, 1980; Bassett, 1986).

Local labour markets and the significance of space

The prospects for establishing the management–labour innovations we have described are not unaffected by spatial considerations. While they are obviously affected by spatial variations in supply of technical skills, as in the contrast between South Wales and the south-east of England, the social characteristics of labour markets are also important, at least as they are seen in management's eyes.

Although labour market differentiation would appear to offer the strongest contrasts at the regional level, managers were well aware of the contrasts *within* South Wales, especially between the valley communities of the coalfield and the coastal belt to the south and east, where managers tended to live. The latter areas have long been the most-favoured locations for private capital, not just for reasons of access (M4 etc.) but for *social* reasons. Companies in this area frequently emphasized its socially differentiated character compared with the more homogeneous class composition of the coalfield. Absenteeism was noted to be several percentage points lower than on the coalfield and adequate supervisory and clerical labour were said to be more plentiful.

One manager of a Japanese plant reeled off a list of rates of absenteeism in plants on and off the coalfield, in south-east England, Barcelona and Japan! Even if such rates have more to do with the management–labour relations in the plants in such places than with the social character of localities themselves, such an awareness is itself interesting. It is this area which is the most frequently 'sold' by the Welsh Development Agency (WDA) in its attempt to eradicate Welsh industrial history from the thinking of possible entrants into the region. The most obvious indication of this conscious emphasis of spatial differentiation within South Wales is the WDA's 'golden triangle' in Gwent (see figure 10.1), which is considered to be more attuned to the specifications of high tech than are the coalfield communities. In one case, community spirit was said to be a threat in the valleys, for the high visibility of a major employer on the coalfield would lead to excessive scrutiny. The triangle has not only the advantage of access to England but better housing and labour markets distinct from that of the coalfield. In other words, although in many respects South Wales still constitutes the most proletarian region of the UK, post-war transformations have fashioned 'locally distinctive class structures' (Cooke, 1981) which, though not wholly new, are increasingly being utilized as a resource by these recent entrants.

These managerial perceptions of socio-spatial differentiation *within* the region were not restricted to new entrants. One of the major indigenous companies, with its traditional facilities on the coalfield, is now following a policy of product diversification in which its new plants are located in the triangle or further east in the M4 corridor itself. Formerly, this company had expanded via *in situ* growth, but this had unwittingly produced the *social* diseconomies associated with the 'mass collective worker syndrome'. Now, having absorbed these lessons, it attaches great significance to small plant size as a means of securing a more motivated, flexible and manageable workforce.[14] In this case, the benefits of small plant size were combined with the perceived attractions of new labour markets in the triangle. Significantly, this company, in which the mechanically oriented AUEW is the numerically dominant union in its (multi-union) coalfield plants, has managed to establish single-union status with the EETPU in its new locations. Similarly, a small new technology firm originating in South Wales now has seven of its nine sites across the Bristol Channel in Avon, largely in order to escape what it saw as 'traditional' attitudes to work.

The pro-growth coalition in Wales

The growth of the electronics industry in South Wales has not been the spontaneous result of locational decisions on the part of capital; rather it has been aided and abetted by the strenuous efforts of a relatively strong pro-growth coalition. With its development area status stretching back to the 1930s, South Wales has for decades looked to inward investment from the UK and abroad as a means of compensating for the decline of its traditional coal and steel industries. A long-established regional problem, combined with a growing Welsh national consciousness, were the main reasons for the creation of the WDA in 1976. The WDA is the institutional core of the pro-growth coalition in Wales, along with the Welsh Office, the Wales TUC and the Wales Confederation of British Industry (CBI), all of which are heavily committed to inward investment. By the standards of the English development areas, Wales appears to have a stronger institutional basis for selling itself to prospective inward investors. Soliciting inward investment is the primary goal of this coalition and its interests are defined in such a way that other regions become the enemy in a territorially based battle for scarce mobile capital.

At the outset it was thought that the WDA would play a catalytic role as an industrial investment agency. But this function quickly became subordinated to what are now its principal functions: provision of industrial estates and customized factory units, fostering indigenous technological innovation, land reclamation and the promotion of Wales as an investment site. Because regional development grants are not sufficient to differentiate South Wales from other similarly depressed regions, the WDA has made considerable efforts to try to create a 'technology-friendly' image. As we have seen this involves an attempt to uncouple the region from its traditional associations with coal and steel, images which are thought to be repellant to high-tech industry. To this end the WDA tends to project Wales as a mixture of rural idyll and cultural oasis. But the main theme of its marketing campaigns is the changing character of the Welsh workforce, a theme which attests to its fears about the traditional social image of South Wales. In one campaign, which suggested a transition from black-faced miner to clean-coated technician, a newspaper advertisement proclaimed:

> The face of Welsh industry has changed dramatically in the past few years. So, indeed, has the face of our workforce. Because most of what we produce these days comes from above ground, rather than below it . . . Which has to be a change for the better.

In a follow up campaign, entitled 'as a workforce the Welsh are anything but striking', it claimed:

> Hand in hand with the silicon chips and fibre optics a new attitude has appeared . . . Many companies now form their agreements with one union and one only. Which doesn't mean there are no disputes. But they do get settled without any of the paralytic seizures of full-scale industrial action.

A flexible strike-free workforce is the main message which the WDA transmits to would-be investors, especially to those from Japan and the USA. What is interesting about the WDA's investment drive is that, publicly at least, it does not try to promote non-unionism, unlike the Tory government's inward investment campaigns. The fact that the WDA does not push an anti-union line stems from its desire to avoid antagonizing the Wales TUC, on whom it partly depends as a co-sponsor of the 'new attitude' in industrial relations.

The Welsh Office injects a degree of political clout into this pro-growth coalition. It may be a pale substitute for a full-blooded state apparatus but it does possess some administrative autonomy vis-à-vis central government. For instance, it is able to negotiate and award regional development grants and other forms of selective assistance to incoming firms, thus making Wales a single port of call in what would otherwise be a procedural labyrinth. Under its Tory political masters the Welsh Office has become a high-profile champion of the cause of high-tech growth, propagating the idea that Wales is now well placed to emulate the 'rags-to-riches' experience of Greater Boston, one of the major US centres of information technology (IT) (Edwards, 1984). Of course, little is said about the critical social and economic differences between the two, such as the fact that Wales is a branch-plant economy, while Greater Boston houses a dense array of HQs and R&D facilities etc. (Morgan, 1987b).

Finally, we should not forget the role played by the Wales TUC. Dramatic changes in its affiliated membership have meant that it is now far less of a voice for trade unions in the coal and steel industries: the National Union of Mineworkers (NUM) and the Iron and Steel Trades Confederation (ISTC) are today dwarfed by the TGWU, the AUEW, the GMBATU and the EETPU, the four largest industrial unions affiliated to the Wales TUC. Inward investment is a top priority of the Wales TUC and here it cooperates very closely with the WDA and the Welsh Office on the question of labour relations. As it says:

> We believe that the record of the Wales TUC and our affiliated unions is exemplary when compared with other parts of the UK. Certainly the role that we have had to play is to ensure the existence and the maintenance of orderly industrial relations and the elimination, as far as possible, of disputes. (Wales TUC, 1984).

Unlike the WDA and the Welsh Office, however, the Wales TUC is vulnerable to pressures from 'below', from its affiliated unions. Some of these have developed a more critical view of inward investment because they feel that some firms are playing one union off against another in the search for single-union, no-strike agreements. In the case of Orion, a Japanese consumer electronics firm, the local management tried to reach agreement with the right-wing EETPU even though EETPU members were in a minority in the plant. As a result other unions feel that the Wales TUC should be adopting a more critical stance on inward investment, to ensure that jobs are not bought at any price. Given the different interests involved it is not surprising to find that this pro-growth coalition is more brittle than it looks.

Conclusions

The implications of the above for the interaction between industry and locality are complex. In some respects the legacy of earlier industrial development in this allegedly radical region has not presented as much of an obstacle as might have been expected, at least to the new entrants, given the younger composition of the workforce, the different labour and locational needs of the new industry and the enabling effects of the recession. Nevertheless, virtually all the firms found it difficult to recruit high-status labour in this proletarian region. Even the new entrants have mostly had to adapt to a unionized environment – certainly an effect of the region's history – and the minority which have not have had to provide better work conditions to keep unions at bay. Here, the speed of growth of the new entrants, plus their possession of greenfield sites, have also acted as a solvent of potential problems in establishing new work practices.

However, while the new practices that we identified have been adapted to the unionized context of South Wales, they are far from unique to the region or to electronics: trade unionists at a Wales TUC meeting confirmed our findings in other sectors. A survey by Income Data Services of four innovative firms in the brewing, confectionary, tobacco and automobile components industries also found similar results (Income Data Services, 1984). The car industry in particular has seen the most publicized cases of the remaking of management–labour relations. For example, prior to 1977, British Leyland (BL) cars had 58 bargaining units for manual workers alone, while the 'mutuality system' allowed shop stewards a veto over new working practices. Among the significant changes introduced since 1977 – aside from mass redundancies and the abolition of mutuality – have been a wholly new group-wide pay structure and the circumvention of the shop steward system as a mediator between management and workers: the BL foreman was now 'a mini-managing director over his own shopfloor'.[15] Stories such as these are now becoming common in other countries too, as far afield as the USA and Australia.[16]

The changes that we have described are ripe for misinterpretation. The Right tends to see the latest developments, including the losses suffered by labour, as part of a generalized solution to current economic problems. At the same time, the far Left either denies that any changes in the nature of work under capitalism could be progressive, unless they were hard-won victories for labour, or else it interprets any admission of such changes as a denial of capitalism's inherent contradictions. The Right's view derives from a failure to consider capitalism as a *system* as well as blindness to the contradictory interests of capital and labour, with the result that the *special* conditions of the success of the strongest capitals in the newest sectors are presented as the *general* conditions for the simultaneous success of all capitals and their workers. The interdependence between success and failure in capitalist competition is denied and the general advance in the development of the forces of production is fraudulently presented as proof that all workers will benefit by allowing capital a free rein.

It is difficult to know how far the changes are progressive for labour or how durable they will prove. On the first question, it must be remembered that the old management–labour practices were neither optimal from capital's point of view nor particularly advantageous from labour's standpoint. The traditional shopfloor power was and is restricted to particular groups of workers, usually skilled, male and white, and often maintained at the expense of other groups.[17] Moreover, since the other side of the coin was the weakness of British manufacturing capital, what was gained in shopfloor control was often lost in job security and pay – which is not to say that ceding this control would guarantee gains in the latter respects. If there is a strong recovery labour is unlikely to remain in its present enfeebled state, but this is unlikely to reverse all the changes in management–labour practices, for they are not all defeats and the previous era was far from a golden age.

As we have seen, trade unions continually have to adapt (not least *culturally*) to keep up with structural change in industry, to colonize new sectors and to avoid being restricted both in fact and imagery to sectors of diminishing importance. High-tech firms represent a challenge to the trade union movement as did white collar employment in the sixties and the 'new industries' in the thirties. It is understandable that unions should feel that concessions are worth making to gain a foothold in firms which are vigorous and relatively secure, even though such concessions may then be extracted by capital in firms which are weak and lack these attractions. It is premature to say whether the gains made by the EEPTU are really footholds leading to strong unionization or disastrous compromises of principles fought for by previous generations. Corporatist top-down deals with firms may provide an entry and circumvent 'grass-roots' initiatives, but whether they can be consolidated in the long run depends on the grass roots: any strategy is likely to involve 'sailing close to the wind'. Unfortunately, at the moment unions are competing rather than uniting to face the challenge.[18]

11

The Electronics Industry in the M4 Corridor

Despite its tag of 'Britain's Silicon Valley', the M4 corridor includes only a small proportion of Britain's electronics jobs – between 7 and 10 per cent. Berkshire contains the main concentration, where, depending on how the industry is defined, they amount to between 5.5 and 10 per cent of total employment.[1] Estimates of the size of the industry in the area also vary widely because the corridor has no

Table 11.1   Electronics employment in the M4 corridor

	Number of employees		Percentage change
	1971	1981	1971–81
Avon	894	1,979	+120
Berkshire[a]	9,227	17,267	+87
North Hampshire[b]	1,725	3,246	+88
South Oxfordshire[c]	221	174	−11
North Wiltshire[d]	9,241	3,169	−65
Total M4	21,308	25,835	+21
Great Britain	405,037	372,501	−8

a. Berkshire Country Council's estimate for 1984 is 30,000.
b. Aldershot, Alton, Basingstoke, Farnborough and Fleet.
c. Abingdon, Didcot, Wallingford and Wantage.
d. Chippenham, Corsham, Devizes, Melksham and Swindon (including Marlborough).

Source: Census of Employment (ER II): MLH 363–367.

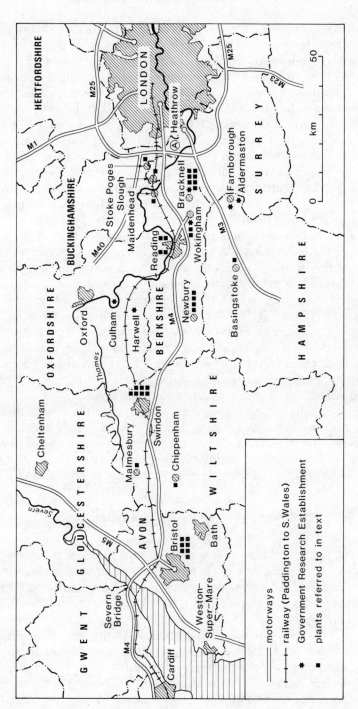

Figure 11.1 The M4 corridor

recognized boundaries. We have included all of Berkshire and Avon, the northern halves of Wiltshire and Hampshire and the southern part of Oxfordshire (table 11.1). On this basis, there were 25,800 employed in electronics in the corridor in 1981, according to Department of Employment data. If Berkshire County Council's estimates are reasonable, then the total may be nearer 40,000. At any rate, on the former figures, electronics jobs in the corridor have increased by 21 per cent since 1971, contrary to national trends, although there have been losses of employment in the corridor too, notably in Swindon.

As in South Wales, there are marked internal contrasts in the electronics industry between old British firms such as British Aerospace or GEC, usually having a strong military involvement, and new foreign, mostly American, firms plus a sprinkling of small new British firms. However, unlike South Wales, there are few large manufacturing plants and large numbers of small firms, reflecting the front-end character of the M4 industry and the more entrepreneurial nature of the area. A high proportion of employment (66 per cent, as against about 24 per cent in South Wales) is in the two electronics sectors with the most consistent employment growth and the largest proportions of managerial and professional labour – computer and electronic capital goods. As one would expect from the corridor's socio-economic character, a much higher proportion of firms are small than in South Wales, though the majority of jobs are in firms with over 100 employees.[2]

In this chapter we shall examine a selection of plants in their causal contexts in order to assess their performance and immediate prospects. They are drawn from a sample of 34 firms, some of them having more than one site in the corridor, and covering almost all the major employers, plus a small number of innovative small firms.[3]

The computer industry

Computer firms, including software houses, account for over 30 per cent of electronics employment in the corridor. The range of products and services covered is wide and diverse, many firms supplying niche markets. Yet, despite this diversity, it is striking just how far these different activities correspond to the changing structure of the industry at the global level, as described in chapter 5. All the firms visited had to some degree been affected by the general changes in the industry: by the shift from a technology-driven industry to a market-driven one[4] with less technically sophisticated customers than hitherto, by the emergence of minis and micros, networking and distributed data processing or, more generally, by the rise of information technology (IT) and the increase in relative importance of software, marketing and customer service. Though not physically present in the M4 corridor (though quite near at Basingstoke), the influence of IBM was evident to some extent in most of the surveyed firms not only as a competitor but as a conditioning environment, setting the parameters of their behaviour. In response to this 'unity in diversity' we have decided to take all the computer firms (including software) together, on the grounds that although this is quite a diverse group it is nevertheless a causal group, albeit a large one, for its internal diversity is not chaotic but strongly structured by interdependences and competitive effects between different parts of the industry.

This influence of the changing structure of the industry is most apparent in firm A. Despite being the largest British-owned firm it is small in comparison with IBM and other leading American companies. Set up in 1968, it has four sites in the corridor and others north of London, in London itself and in the north-west and Scotland, plus a sprinkling of regional service centres. Many of the sites were inherited from the nine separate companies from which it was formed. A has been affected by most of the recent changes in the nature of computer products and by the increased internationalization of the industry with its attendant intensified competition. For years it had tried to be a 'mini-IBM', concentrating on mainframes but attempting to make and sell a full range of products. With the aid of preferential government purchasing, in the days before minis, micros and distributed data processing became established it achieved some success as a technology-driven company, doing its own basic R&D and selling what were often very advanced machines to technically sophisticated customers. However, it lacked economies of scale in production and marketing and could not match the power of IBM's sales and customer support.

Now the picture has changed radically. At first, A tried to meet the growing diversity of the market by making its own versions of new products. In doing this it felt even more the disadvantages of its small size and consequent lack of economies of scale and marketing power. Preferential state purchasing also had to end with Britain's entry into the EEC, and in any cast IT markets have been liberalized. So, in 1981, after a financial crisis, a major programme of restructuring was begun in which the company moved decisively away from being a computer manufacturer to what the managing director described as a vendor of 'high value-added systems' and 'solutions to customers' problems'. In other words, the company was to give up trying to be a mini-IBM, to cut back on in-house manufacture and to make up the rest of the product range necessary for providing 'total solutions' through collaboration with other companies, buying in their products and 'adding value' to them. Deals have been made with a variety of firms: a Japanese firm to make key chips;[5] a North American telecommunications firm to supply private exchanges; and various other British computer firms and retailers for the production and marketing of particular products.[6]

The changes have involved heavy redundancies: total employment in the company fell from 34,000 in 1979 to 25,654 in 1981, and to 15,500 in 1984 (22,000 world-wide). Now, for the first time, its employment in Britain has fallen below that of IBM (17,000 employees). Something of the weakness of A's position is indicated by the fact that IBM UK's sales in the UK were nearly double those of A world-wide in 1984. Not surprisingly in view of the direction of the changes, most of the redundancies were in manufacturing, in which employment fell from 10,000 to 4,000. Not all of these cuts were due to reduced in-house manufacture: a small part in recent years has been a result of the introduction of highly automated 'flexible manufacturing systems' as well as the usual effects of microelectronics in reducing component counts. Proportionately fewer of the redundancies were felt directly in the M4 corridor as manufacturing was located outside the area, particularly in the north-west.

Small increases in employment have occurred in those activities essential to 'selling solutions' – marketing, software and development (for customizing products) – and these have been historically concentrated in the M4 corridor, in accordance with the usual spatial division of labour. Given the difference in skills between these activities and manufacturing there has been no attempt to redeploy redundant workers. Marketing staff have achieved higher status and in response to the needs of technically unsophisticated users the company has tried to improve liaison between marketing and product development – activities formerly separated by a bureaucratic and social divide. Like many computer firms, A has also reorganized itself on a market basis instead of a product or activity basis so as to facilitate contact with customers requiring 'solutions'. Geographical shifts resembling 'musical chairs' have characterized the restructuring process. Some sites have closed altogether and departments have been moved between sites without prior consultation of workers, with the result that some of the most skilled have left, often to join foreign firms better able to use their skills.

This change in role towards that of system integrator has, in management's eyes, given the company more realistic aims. However, like all the leading European companies, A faces difficult choices in response to the impossibility of continuing in its traditional role, and the alternatives are highly risky. In partly becoming a system integrator, A is putting itself in a more reactive role in the market, reducing long-term planning and failing to use some of the R&D expertise it built up as a committed manufacturer. As some union members argued, A is losing its ability to innovate. Yet outside of some wider state-backed scheme, it is hard to see what else A could do. Much will depend on 'the battle of standards' in operating systems and networking, as this will affect the extent to which A is exposed to competition from larger American and Japanese firms. With some successes in its new role, especially in computer systems for retailing, the position of the company is almost certainly stronger than it would have been without the restructuring. Yet its profits are only modest and in employment terms it has been a failure, though less so in the M4 corridor than elsewhere.

Several M4 firms addressed minicomputer markets, whether by selling their own models or as system integrators adapting and selling other firms' products. Company B is a large American-owned company with 6,000 employees in Britain, a third of them at three sites in the M4 corridor, one of them established in 1964. It also has an assembly and test plant in Scotland and two manufacturing plants in Eire and one in West Germany. None of these factories serves an overspill function and generally each of B's factories serves the whole world market. The same is true of its R&D establishments, most of which specialize in a single function for the whole company and some of which are located in Europe. The importance of supplier–user contact is illustrated by its 14 sales and service centres and 50 'remote service locations'. The M4 plants include service and sales work, UK headquarters functions, specialized R&D, a limited amount of engineering, and customer and in-house training.

As the leading specialist in minicomputers, it has a different profile from company A, though again it has had to adapt to similar changes. Given its size, its problem was not lack of economies of scale but the development of the micro market and the shift to less technically sophisticated customers. The business had always been centred on hardware, though its success was based on reliability and customer support rather than being a technical leader, but now it had been obliged to try to enter the micro market and to sell to uninitiated customers. However, this meant increasing its involvement in software, which it was reluctant to do, and in marketing, which involved a reorientation of the company and its traditional skills. The apparent strategy appears to be one of compromise in which a large proportion of sales go to system integrators who reconfigure the systems, adding value to them for particular types of customers. This allows greater standardization and hence cost-lowering than would be the case if they were to sell all their output direct to end-users in customized form. Much of the software work is subcontracted to some 50 software houses in the locality. While such a compromise allows the company to continue to play to its traditional strengths, many of these collaborators and subcontractors are of course actual or potential competitors and so the strategy has the possible disadvantage (from the company's point of view) of encouraging them.

B has not had any redundancies though it has made extensive use of contract clerical workers, systems analysts and programmers while informally offering lifetime employment to core workers. Nevertheless, it has had difficulty in redeploying many field service engineers whose jobs have become less necessary with the development of remote diagnostics and more reliable systems. While problems with its micro venture caused the company's growth to falter briefly in 1983, it has recently grown at twice the industry average rate, largely because it has taken a lead in advanced networking and in offering compatibility between products. As a result, it has announced a further 1,000 jobs in R&D and sales and marketing support, 600 of them in the corridor. Once again, this demonstrates how company performance is not merely based on manufacturing efficiency; the quality of the product, marketing and after-sales service often matters more.

Company C is also American and involved in the minicomputer market. Established in 1938, it has its origins in the market for sophisticated electronic test equipment and this still makes up over a third of its business. Having diversified into calculators and computers it found that technical developments in its various divisions showed a strong convergence around computing, especially software. More recently, the development of distributed data processing and IT has encouraged C to move away from an organizational form based on discrete products to one based on systems for particular markets. While this has called for greater integration of activities, the firm has retained its policy of decentralized organization based on small units by making full in-house use of the most advanced IT products to coordinate these units (see chapter 12). With sales growth rates of 30 per cent per annum, and for many years employment growth at 20 per cent per annum, it has made the transition from serving technical customers to serving commercial customers more successfully than some of its rivals and has been especially successful in office automation.

In 1984 it had 84,000 employees world-wide in 71 countries, with 3,100 in the UK, a Scottish plant with over 1,000 employees manufacturing and marketing communications equipment, 23 sales branches, and 1,500 employees at five locations in the M4 corridor.

One of these employs over 200 in assembling and marketing computer peripherals. Another site has 290 employed in a world-wide software R&D centre for office automation, together with customer support and UK headquarters functions, while a recently opened site houses C's European R&D centre. While over half of C's manufacturing sites do some limited development work, adapting US-designed products to local national markets (again, contrary to the stereotypes of spatial divisions of labour), these two R&D centres are the first ones outside the USA to do more fundamental research. Among the reasons for their existence is C's interest in European, especially British, research perspectives in computing and artificial intelligence, and the UK was seen as a cheap source of graduate software professionals. While many peculiarities of European markets only require adaptation of American products, some, such as the European networking open systems interconnection (OSI) standard have more fundamental implications which cannot be met simply in this way.

However, despite its high growth rates and reputation for advanced management methods – of which more in the next chapter – not even this company has proved immune to the present computer slump and, while no redundancies have been announced, a 5 per cent pay cut was introduced in 1985 in the USA and Britain. Rapid growth in this market requires time economies in innovation itself and some American business analysts have noted that the company has postponed its major new product launch several times now so that investors and customers are in danger of turning to its competitors. This is despite the company's policy of committing 10 per cent of annual turnover to R&D. Once again, a rapidly growing industry may be far from a low-risk environment, even for leading companies.

In this kind of market, quite small firms quickly go international. Company D, another US-owned minicomputer firm with only 1,000 employees world-wide, and just 50 in the M4 corridor, specializes in the markets for industrial control systems and computer-to-computer communications and found demand for the latter to be strong in Europe, particularly with the liberalization of telecommunications in Britain. As one of the first firms in these markets, D enjoyed some success in winning large contracts in the City of London and with airlines. It located sales offices in the corridor in the late seventies and then established a small manufacturing plant (50 employees) there, plus its European headquarters, which covers marketing, sales and software development. Like B, it tried to diversify into making microcomputers in 1983 but with disastrous results. These problems came on top of a slow-down in industrial demand for computers and a failure to develop new products. Following this, manufacturing – or rather assembly and testing – was transferred from the M4 corridor to the Irish Republic, a much cheaper location because of its regional incentives and lower wage rates.

Another small but more successful firm, E, is European-owned and has its central European headquarters in the corridor.[7] It began as a producer of 'superminicomputers' chiefly for scientific applications but, in accordance with the changing nature of the industry, it now describes itself as selling integrated information systems covering hardware, software and networking products. As the small scientific markets have become saturated, and as hardware has become more in the nature of a black-box commodity with low margins, it has tried to shift into faster-growing activities such as office automation. The latter kinds of market need far more marketing, software development, documentation and customer education and support, and keener pricing too, although E also sells to systems integrators who perform some of these functions. Given its early involvement in scientific markets requiring minimal support, E has had further to move in this direction than some larger rivals. This change, coupled with the need to break out of the limits of its tiny domestic market, has prompted a strong emphasis on exports and foreign direct investment including collaboration with foreign companies such as a British defence company. For a small firm, quite a number of key hardware and software products are made in-house, chiefly in the home country, but many components, subassemblies and peripherals are bought in from big companies and, like many computer firms, it makes frequent use of leading software houses as subcontractors.

A purer example of a systems integrator is company F – actually a group of three small firms – all located in the corridor and operating as technically highly specialized systems integrators.[8] They buy in chiefly American-made hardware and 'reconfigure' it for specialized functions such as image processing and sell mainly to big companies and scientific establishments. Production is mostly in projects or very small batches. While the companies have had high exports, their growth has been steady rather than rapid. Again, it makes extensive use of local subcontractors both for simple components and subassemblies and for sophisticated software applications.

Company G actually closed down its M4 office automation plant in 1986. Set up in 1979 as an offshoot of a prestigious British software house, it produced word processors and was beginning to move into local area networks for linking computers and workstations. As is typical in small European computer firms, hardware was partly bought in (mostly from the USA and Japan), partly made in-house and sold mostly to other system integrators or to larger companies for marketing, rather than direct to customers. Given the nature of the business and the extensive use of automation, it had as many professional engineering staff as operators and included R&D, software and sales and marketing on site.

Again, typically for many such firms, its rise and fall was rapid. In 1983, it led in word processor sales in Britain with 26 per cent of the market, against much larger and older competitors like Wang, DEC, Olivetti and IBM. This success was largely thanks to a major contract from British Telecom and others from a leading British computer firm. Employment reached a peak of 400 in June 1985. Then the computer slump struck, coupled with the entry of IBM into the market, and by October of the same year company G had begun its unsuccessful search for a buyer for the subsidiary, and workers were being redeployed or made redundant. While the venture had many advantages, such as expertise in software and office automation and extensive use of computer-aided design, automated assembly and testing, these were not enough. In particular, it was too small to achieve economies of scale; though it avoided marketing problems by selling much of its output via systems integrators, it lacked a cheap manufacturing labour force and its managerial strength in highly specialized software – which continues to support the remainder of the company – was not appropriate for the very different skills needed for running a low-cost manufacturing operation.

Whereas G's failure had much to do with the entry of IBM into its market, company H, a software firm, has found that IBM's entry into microcomputers and linked mainframe-micro systems has actually created opportunities for it in the shape of a new market niche.[9] If micros are to run the same data-processing programs as the mainframes to which they are linked, they have to be able to use the same sophisticated programming language as the latter, and this need has provided H with its main market. Generally, many software firms were aided by the arrival of the IBM PC and the resulting shake-out because it stabilized the micro market, though H was unique in filling this particular niche. The niche is quite large, however, as the products can be used on most computers in most countries. Like many software companies, it also concentrates on so-called software tools, which facilitate software design and testing. It sells primarily to computer manufacturers whose products are enhanced by the incorporation of these tools in their packages. Relationships with manufacturers are therefore close and frequently the latter send samples of their computers to H for 'fitting' with their products. For this reason and generally to facilitate 'learning by using', H is a large user of not only its own products but those of its customers.

Like many European IT firms, it found it essential to set up a branch in the USA (in Silicon Valley) at an early stage, not only as an entry point into the vast US market, but as a 'listening post'. The US site has similar functions to the M4 site and the two are linked by satellite. Unusually for such firms, it has had a fair success in penetrating overseas markets with 58 per cent of sales in the USA, 27 per cent in Japan and 13 per cent in Britain. However, it too has been affected by the computer slump and has announced major redundancies.

Semiconductor firms

Among the most prestigious firms in the corridor are branches of two large American semiconductor companies, I and J.[10] Both house mainly front-end activities and, in J's case, nearly 60 per cent of the workforce are professionals. J has primarily been committed to the fiercely competitive commodity end of the market, I to more design-intensive chips. While both have suffered in the chip slump, not surprisingly the former has had most difficulty given its greater exposure to Japanese competition. Consequently, it has had to make lay-offs at its Scottish fabrication plant and to introduce short-time working at all its sites and, world-wide, has shut down 25 per cent of its capacity. I lacks any European fabrication facilities but had cut 20 per cent of its production capacity in 1985. It has also made pay cuts in a previous slump and admitted to being overstaffed.

In this context both are trying to shift their centres of gravity towards more complex products in which price competition is less fierce and where customers are less free to shop around, though I is also reviewing its policy of allowing other firms to do the bulk of its standardized manufacturing. J is moving into 'systems products' which might be used as subassemblies or development tools by IT equipment manufacturers, and I is cautiously trying to move into more advanced products and semi-custom chips, though without forfeiting its massive stake in commodity components which provide most of its income. Gaining the custom of big computer and telecommunications companies – their key market for these more complex products – is also tough, and requires greater emphasis on the cultivation of collaborative supplier–user relationships in marketing, development and support, but once established they are less likely to be vulnerable to price competition than with 'commodity' chips.

A third, and very different, semiconductor firm, K, is British-owned, with about 690 employees at its M4 site. Some fabrication is done here though an additional facility is being added in south-west

England, while R&D work is carried out at three centres, one of them also in the corridor. All but very low-volume assembly work is done in the Far East. K is uncomfortably positioned between its traditional small British defence and telecommunications markets, which it has outgrown, and larger international markets which, as yet, it lacks the power and size to penetrate. And, while its specialized markets have so far largely insulated it from the chip slumps, it has also missed out on the booms. If it is to reposition itself in the market as it hopes, K has to cross the classic barrier which the British electronics industry has allowed to encircle itself, namely that between low-volume, slow-growth, specialized production for small state-sponsored niche markets and the quite different approach and skills needed for volume production and marketing in a more open, competitive and international market.

K's strategy is not to move into 'commodity products' (like firm D in South Wales) but to gain access to rather larger niches in civilian markets than hitherto, particularly in telecommunications. To this end it is starting a bold venture in a new kind of semi-custom chips, and trying to get into the US market. To stand any chance of selling to big American systems companies it has to establish close relationships with them and gain credibility as a reliable high-quality producer. To this end it has acquired two American systems companies (though one of these is now in trouble) and it is seeking to buy a US fabrication plant. To make much impression on the rapidly expanding markets it will have to grow at up to 50 per cent per year and raise the proportion of its sales going to the USA from their present low level of a quarter. To do all these things it will have to develop both a more aggressive form of marketing and the proficiency in process engineering required to make volume production a success.[11]

Firms specializing in producing semi-custom chips for outside customers set up 'design centres', where engineers from customers' firms can go to design their circuits with the aid of the suppliers' design tools and guidance. While the above firms are setting these up as additions to existing activities, one US company, L, was set up to specialize wholly in semi-custom chips, and it has one of two European design centres and its European headquarters in the corridor. It has wafer fabrication plants in the USA and also one under construction in West Germany. Established in 1981, the company had 1,400 employees world-wide in 1984 with 38 in the M4 corridor. The company is highly research intensive and has devoted most of its efforts to developing user-friendly design tools and rapid turnaround of customers' designs. (One of the advantages of concentrating on these features is that it makes the customer bear many of the labour costs, of scarce engineers and expensive supplier-buyer interaction, involved in design.) While L has been highly successful in the world market and has been less affected by the chip slump than larger producers of standard chips, its growth in Europe has been slower than in the USA and Japan because of the relatively backward state of the user firms here.

In their different ways, then, the performance of these four semiconductor firms in the M4 corridor reflects both the general state of the industry and the structure of uneven development in which Britain and the corridor are situated.

Telecommunications

Although quite diverse in size, ownership and products, all the M4 telecommunications firms have been affected in some way by the impact of the chip and IT and by the impact of the liberalization of the telecommunications industry.[12] They are also likely to be affected in the near future by the expected shake-out in the industry resulting from the imminent saturation of parts of the market, as too many firms are offering products such as public exchange and private automatic branch exchange (PABX) equipment in relation to current demand. The differences in their behaviour result substantially from their different positions with respect to these changes, i.e. in terms of how advanced they are in microelectronics, and whether they were 'protected insiders' in the industry during its state-regulated days or whether they are newcomers – whether in the form of new start-ups or inward investors.

Company M, with 150 employees in the corridor, is without doubt a technological leader, having been the first in the world to complete a family of fully digital equipment in 1976 and having great strength in the two foundational technologies – semiconductors and software. As a vertically integrated company (in fact as well as name) extending back into microelectronics production, it has the advantage of being able to capitalize on technological synergies.

M has grown extremely rapidly: turnover has increased more than fourfold in the last ten years and employment world-wide had nearly doubled, though, thanks to the state-regulated nature of most national telecommunications markets, its overseas investments are limited. While the company has 41 manufacturing plants and 37 R&D centres in North America, it has only two manufacturing facilities in Europe – one in Ireland, the other north of London. In 1981 it set up its first overseas R&D laboratory in the M4 corridor, having previously only had a sales office there. After many years of concentrating on the North American market (equal to half the world market), it is now turning its attention to Europe, Africa, the Middle East and India, using the M4 site as a base. The UK was favoured primarily because, thanks to liberalization, it was the most advanced market in commercial terms (M's UK sales equalling those of the rest of Europe as a whole) though in technical terms the UK and the rest of Europe were and are far behind North America. Given the semi-customized nature of many telecommunications products and M's policy of making R&D market oriented, the company decided to open its first European R&D centre in Britain. It has already licensed or sold PABXs, public exchanges and other advanced equipment in the UK. At the time of interview most of the products were imported but the company planned to open a manufacturing plant, also in the M4 corridor. Scale economies relative to the size of European national markets were such that each manufacturing plant would serve several countries with particular products.

Because of its size and vertical integration, M has little need of local sourcing and its chief local impact is likely to be in driving up demand for already scarce telecommunications and computing skills. Recruitment is likely to be cautious however, for it has a reputation for 'running lean', having achieved the same turnover as Ericsson, the Swedish manufacturer, with half its employment. Given its size, wide product range, aggressive international marketing and technological leadership (deriving from an R&D budget far in excess of British telecommunications firms) it must constitute a severe threat to established British telecommunications firms such as GEC, STC and Plessey, as British Telecom (BT) liberalizes its purchasing policies and as markets saturate. While there is no doubt that M will be not only a survivor but a beneficiary in the coming industry shake-out (it has already mopped up abandoned R&D capacity – and presumably staff – from one major British firm as it disinvests), the employment it creates in Britain is certain to be far less than the losses in rival firms which it will indirectly precipitate.

If M typifies the dilemmas of attracting advanced foreign companies into Britain, so too, in a very different way, does company N.[13] An established firm dating back to 1919 and taken over by a European multinational in 1967, it was then operating in a highly regulated state-controlled market and was technologically a follower rather than a leader. Unusually it foresaw that neither the market conditions nor the electromechanical technology would prevail in the long run. It therefore embarked, in the early seventies, on a major programme of microelectronics R&D, thereby becoming for a while one of the more technically advanced producers in the

UK, making simple telephones, exchanges and peripheral equipment. Between 1978 and 1982 output quadrupled and employment increased by a third. However, much of this was due to a rush of purchasing by BT, and since that time turnover has dropped and liberalization has brought in more advanced firms, thereby intensifying competition. As a result of this, plus the usual effects of reduced component count in systems manufacture, employment was cut in 1984 by about 12 per cent at its M4 design and development site and its two Scottish manufacturing plants and finally, in 1986, the M4 site was closed altogether and its operations were transferred to Scotland.

While liberalization had also created niches for small firms, some of them located in the corridor, their prospects of growing were limited. In the emerging context of oversupply and probable shake-out, all the small and medium-sized firms were aware of the danger of entering a downward spiral of declining market share, lack of funds to amortize R&D, outdated products, declining market share and so on, leading to ultimate failure.

Military electronics

The prominence of the arms industry in Britain's sunbelt is particularly clear in the corridor; Lovering estimates that it accounts for 12 per cent of manufacturing employment in the south-west region and probably 2 or 3 times that in Bristol. While in the latter case this localization has come about partly fortuitously, a major factor in the concentration of firms in and around the corridor has been proximity to government research and defence establishments, such as Aldermaston and the Royal Aircraft Establishment at Farnborough. The procurement relation basically consists of the Ministry of Defence (MoD) providing funds for research together with exacting specifications, followed by close liaison throughout. Over the years a stable network of information linkages has built up between arms suppliers and these establishments, producing considerable locational inertia. Many M4 firms sold small proportions directly or indirectly to military customers. Among those that were primarily oriented to military products were firms O and P.

O has three divisions in the corridor, two of them specializing mainly in electronic equipment and employing 2,500 and 4,960. In addition it has several other sites, most of them located in southern England. One of the M4 plants serves as a sophisticated jobbing shop for other divisions, designing, developing and supplying equipment, while the other specializes in R&D and manufacturing missile systems. Naturally, computerization figures prominently in the latter and the plant has made 16 different types of military computer. While output has grown rapidly, with 1984 sales up 70 per cent on 1980, the firm has begun to feel the pressures of operating under fixed price contracts. In response it has sought new civilian and overseas defence markets and has tried to alter its 'corporate culture' to give a more commercial orientation. However, compared with other electronics firms, it is still relatively protected and enjoys a large amount of work in particularly long contracts, some of them lasting for up to 25 years. By comparision, firms selling minicomputer systems for civilian customers find that, unless projects are completed in under 3 months, markets pass them by.

There was also some small batch production, e.g. for certain missile units, and O was introducing computer-aided design and manufacture expanding the use of computer numerically controlled tools. However, the reason for this was not merely to increase productivity but to improve consistency and quality. Not surprisingly, given the kind of work, at the jobbing shop plant the largest groups in the workforce were scientists and technologists, with technicians following, and 25 per cent were graduates. Similarly, at the other site only 25 per cent were involved in direct production, while 35 per cent and 20 per cent were involved in design engineering and support engineering respectively. As is typical of the sector as a whole, skills were shifting from mechanical to electronic, from analogue to digital, and from hardware to software. While there was some tendency towards jobless growth, continued growth in defence expenditure and the customized nature of production, while inhibited economies of scale, meant that there was still some growth in demand for more specialized skills.

A more modern face of military electronics is presented by company P, a rapidly growing 'middle league' British electronics firm specializing mainly in military equipment for both the MoD and commercial overseas markets and having a noticeably more commercial overseas markets and having a noticeably more commercial orientation than firm O. It has a wide range of products including communications equipment, electronic warfare equipment and radar on the defence side and a range of products on the civilian side including modems (for linking computers), artificial intelligence, cable television, computer-aided design technology and test equipment. Having started in West London, it now has 18,000 employees world-wide and over 120 subsidiaries, many of them in the English sunbelt, including the M4 corridor. Unlike some of the

older defence companies which lack a strong commercial orientation, P has kept units small and targeted them on particular markets.
Most of these units have well below 500 employees and function
as independent profit centres responsible for their own finance,
manufacturing, sales, service and R&D – again contrary to the
stereotypes of spatial divisions of labour. Only long-term, strategic
R&D of potential use for many parts of the company is done by the
parent company. Many of the subsidiaries started off as acquisitions,
and while P has been very successful in buying up firms in trouble
and making them profitable, this has usually involved redundancies
for many of the original employees; for example, in the case of
the acquisition of an old British radar firm, troubled by labour
disputes and lacking a clear commercial strategy, the workforce
was reduced from 1,800 to 55 and the London site closed.

Production types range from project and small batch in defence
to medium volume in cellular telephones. Some small defence
products are also produced in volume for inventory in readiness
for sudden orders by countries at war. This also contrasts with
some of the older, more MoD dependent firms both in terms of
product type and commercial orientation.

Unusually for a British electronics firm, P had some success
in penetrating the US market in which its sales equal those in
the UK, though of course, given the size of the former, market
penetration is small in absolute terms. An important factor in
this respect was its emphasis on providing service back-up for
sales, e.g. enabling one of its subsidiaries, with 119 centres, to
achieve a two-hour maximum response time in the USA for
certain products. Nevertheless its success has been short lived
and the company has run into difficulties in the USA, illustrating
once again the difficulty of penetrating advanced foreign markets.

In general, then, firms in this sector enjoy a relatively protected
status and, while competitive pressures have begun to impinge in
arms production, there is no equivalent to the shake-outs occurring
in the semiconductor, computer systems and telecommunications
sectors. However, where they have attempted to penetrate civilian
markets they have had to adapt their skills and organization
radically and even P, with its numerous product-specific profit
centres, has found this difficult.

Other companies

Certain firms or plants cannot easily be put in a single category,
but understanding them need not be difficult provided allowance

is made for the fact that they operate in several causal contexts. In some cases, the common technology of microelectronics and IT allow a single firm to operate in a wide range of sectors. As this is a common feature in any industry we see no point in trying to disguise the fact by forcing every firm into a single category.

Three major M4 firms illustrate this kind of situation. Q is a subsidiary of a large US multinational which had been operating in the UK for nearly 50 years, producing computer systems and instrument engineering. The M4 branch is part of a division which used to specialize in electromechanical control systems. With the development of microelectronics, electrical engineering gave way to electronics and the sector converged with that of computer systems. This technical convergence plus the lowering of barriers to entry caused by microelectronics has exposed many engineering firms to greater competition, both from established computer firms and systems integrators and from new start-ups. However, this has been less of a problem for Q than for some firms, since it belongs to a group of companies in which computing figures strongly. These origins, plus the presence of a nearby site of the information systems division, have led to an unusually diverse range of activities. This includes manufacturing aerospace, defence and medical equipment, security and control systems, e.g. for oil refineries, servicing a huge existing customer base of installed products, doing software development for business computer systems, providing data-processing and telecommunications services and giving courses on new technology.

Manufacturing has been carried out in an old-established Scottish plant. Employment there has dropped from 1,400 to stabilize at 800. The M4 site specializes in administration, sales, installation and service work; half its 800 employees are field service engineers, many of them working from home. As in company B, with the development of modularized products servicing has been deskilled and this, coupled with the rise of remote diagnostics, has resulted in a surplus of such engineers, though service contracts are a large growth area.

Company R had also moved from electromechanical to electronic and computer engineering, having previously specialized in equipment for railways (e.g. brakes and signals). It now produces electronic components and process control and communication systems for railways, sewage and energy supply, and has partly become a specialist technical systems integrator, buying in and reconfiguring computer hardware, most of it coming from American firms, and combining it with electromechanical equipment. Both R&D and manufacturing are done on site and production includes both project work (e.g. for railway signalling systems) and batch production of components. As usual, customer liaison and after-sales service is very important for the systems products. Employment had fallen sharply from over 4,000 to 2,500, partly as a result of reduction in investment by customers during the recession.

A further firm working in the grey area between electronics and mechanical engineering was S, which had an M4 plant particularly distinguished by its exceptionally advanced automation. With the aid oF flexible manufacturing systems including fully automated materials-handling system, robots and automated assembly lines, all coordinated by computer, the factory makes a wide range of electronic vending machines, plus small radar outfits and electronic test equipment. Thanks to computerization a new product variant enters the system every two hours with minimum delays between batches. Without this degree of automation, producing so many variants in low volumes would be prohibitively expensive.

This firm has been highly successful in penetrating foreign markets and output has grown at rates of between 40 and 60 per cent since it was set up in 1973. Consequently a new, additional plant – basically an overspill plant for producing more of the same products – is under construction nearby in the corridor. Despite its exceptional degree of automation, such high rates of growth have required it to increase employment steadily – e.g. from 480 in 1983 to over 600 in 1984 – and the proportion of the workforce who are managers is high at over a quarter.

In addition to the firms mentioned so far there were a number of small manufacturing plants, mostly involved in making components and small subassemblies and having a higher proportion of operators and back-end functions than most of the larger plants we have discussed. Typical of these was a subsidiary of a Welsh firm (firm C in chapter 9) employing 55 people and making printed circuit boards. 80 per cent of its market of civilian and defence firms were located within a 50-mile radius and it provided a jobbing function, producing small batches at short notice for which proximity to customers was essential. Even small plants need not escape jobless growth; automatic testing and computer numerically controlled drills had enabled nearly a sixfold increase in output without additional operators. It was firms such as these which had the stronger linkages with local suppliers and customers, in some cases this being partly a function of their limited resources for marketing.

Company characteristics and performance: general conclusions

Having looked at the M4 plants in their competitive contexts, we are now in a position to draw some generalizations about dominant characteristics and processes.[14] In this case, the order of presentation reflects the research procedure, for we wanted to avoid making premature generalizations without checking whether common effects arose for the same reasons. For example, without a prior causal analysis of each plant, we might easily have ascribed the widely observed intensification of competition and increased commercial orientation to a single, uniform cause. Nevertheless, despite the diversity of activities and the number of niche suppliers, virtually all the firms related strongly to major trends in their sectors. To be sure, there were many contrasts, but the different behaviours of different firms were often interrelated, producing a picture of unity in diversity, or at least a small number of unities in diversity.

Product technology and markets

All the older firms had been affected by the development of micro-electronics, replacing and supplementing earlier electromechanical technologies, and by the convergence of engineering and computing. This had caused more problems for the British firms than for the more internationalized and advanced foreign firms.

The nature of competition

In the great majority of cases we also saw evidence of increased competition and market orientation. One reason for this general to all sectors: the increased penetration of the UK market by foreign firms, as a result of inward investment. However, there were other reasons, which differed by sector.

In semiconductors it was largely due to the world-wide crises of overproduction, which was making firms shift to new products whose markets were not yet saturated and for which added value was great-er.[15] While these products require little or no after-sales service, the buyers' market for components, particularly integrated circuits (ICs), was obliging them to pay more attention to marketing than hitherto.

In computer firms, competition was increasing as a result of slackening demand and the maturing of former infant industry subsectors. The most dramatic consequences were the closure of firm G – which suffered from problems typical of small firms trying to enter rapidly growing volume markets against bigger firms – and the rationalization of the in-house product range of firm A although, in the latter case, most of the resulting job losses occurred outside the M4 corridor. More striking still was the shift by all the computer firms and systems integrators towards greater market orientation and increased emphasis on software and development work.

Some of these trends were also visible in the telecommunications firms, but here it was liberalization which stepped up the intensity of competition, partly through attracting foreign firms into the UK, forcing indigenous firms to develop a sharper marketing effort. Again, overproduction and the beginnings of a shake-out were apparent.

In the arms industry, a greater commercial orientation was emerging as more competitive forms of contracting by the MoD were beginning to be introduced and as firms tried to diversify into export work and civilian markets in order to grow faster. So, both in telecommunications and defence, firms which had for a long time sought the security of state-protected markets were now having to change direction.

In many cases, perhaps the majority, improving product performance was a more successful way of competing than improving manufacturing efficiency so as to lower price. Given that products such as computers are used as means of production, what matters is not simply their price but their price-performance ratio over a number of years. Hence as one manager in a producer firm in a survey by Davies put it:

> The company's way of making profit is by exploiting talented individuals in the development department rather than sophisticated techniques in the production department. (Davies, 1984)[16]

One of the few cases where process efficiency in manufacturing was the key ingredient of competitiveness was firm S, with its advanced flexible automation.

Activities in the M4 plants

Sectorally, producer electronic industries are much more to the fore in the corridor than in South Wales, largely reflecting the need of IT and defence firms to be near customers, who are also heavily concentrated in southern England (cellular phones and radar for small pleasure boats were virtually the only consumer goods produced by the survey firms). Generally the activities in the area reflect its placement in spatial divisions of labour which are organized on *both* hierarchical and regional-sectoral lines.

Front-end activities such as marketing, development and service work are far more prominent rather than in South Wales; what little production exists is mainly small batch, custom or project work. However, several plants also carry out the more mundane tasks of warehousing and distribution; indeed this was much more in evidence than in South Wales. Non-manufacturing activities predominate in most plants in the corridor; in A, the proportion of manufacturing employment was just over a quarter, in B just over a fifth, in C nearly a third, in P a quarter, and in many, such as E, zero. These proportions reflect several things: first, the prominence of high-status activities in the industry as a whole (e.g. the large proportion of professionals) and the importance of product innovation, marketing and service work; second, the corridor's position at the top of the corporate hierarchical divisions of labour in British firms; and third, the presence of several foreign firms, most of whose manufacturing is carried out elsewhere, particularly in their home countries, and whose M4 activities generally occupy an intermediate position in their corporate hierarchies. A possible fourth cause of the lack of manufacturing, particularly as regards computers and telecommunications, is the large proportion of foreign-made components that British firms buy in.

In software, there is of course no manufacturing labour and most of the work is quite highly skilled. This implies rather different corporate hierarchies than those familiar in manufacturing. Davies cites the example of the British 'service and support' branch of a leading American software company (Digital Research) employing 18 people in the corridor (Davies, 1984). Major products are invented in the USA mostly by one man and are then developed by a team of highly skilled personnel. They are then passed on for 'engineering' – documentation, quality assurance and servicing – again, highly skilled jobs. All of this is done in Silicon Valley. The M4 plant is involved in some development work for particular customers as well as service and support, and 13 of the 18 staff are skilled managerial and professional sales, legal and programming staff. Yet, despite its skill profile, this operation is low down in the corporate hierarchy.

R&D was common but with the emphasis on development and mostly involving not a discrete phase following research on a particular innovation but a continuous process of adaptation and improvement of products (particularly software) already in use, in which development merged into engineering. Given the importance of software and of custom and project work for industrial, commercial and public sector use, this is not surprising. Where foreign firms are concerned, 'R&D' has traditionally meant adaptation of foreign products to suit British standards and markets. However, there was some evidence (e.g. firm C) that more fundamental kinds of R&D were beginning to be introduced in order to respond to standards and markets which are too different for mere adaptation, to tap UK skills, and, as a by-product, to win political approval from the host government, Again, contrary to the stereotypes, several firms had co-located R&D, or at least development, with manufacturing. This fits with the large proportion of custom and project production in the corridor, but it might also possibly be evidence of a trend in large multiplant firms towards a situation in which only long-term strategic research which transcends existing activities and divisions is housed separately.

The M4 plants were clearly of a different kind from those of London, South Wales and Scotland. Yet even with its predominance of front-end activities, the corridor's many leading international firms tended not to be as innovative, nor of such strategic importance or employment or spin-off generating capability, as those of their places of origin. Thus in international terms the corridor occupies an intermediate place in corporate hierarchies.

The performance of the plants and firms

Overall, most of what we saw in the survey was consistent with wider assessments of the UK industry (see chapter 8). British-owned firms tended to be weaker than foreign ones, and while most experienced output growth, there were few unambiguous cases of British firms gaining a significant share of the world market. Secondly, there were many clear examples of the establishment of branches of foreign firms having an import-inducing, rather than an import-substitution, effect. This was particularly clear in the case of recent telecommunications and computer entrants which had not yet established much manufacturing in Britain. Both foreign and British systems integrators relied heavily on imported components, equipment and software. In this respect, IBM, with its balancing of intra-company trade, is the exception. The long-noted practice of British electronics firms of retreating into protected state markets was clear in the survey, though there were also a few cases of attempts to escape these and enter larger and more competitive markets, some of them prompted by reduction of state protection, as in telecommunications, and often involving investments in the US (e.g. firms K and P). However, the risks of entering more competitive markets such as semiconductors, PABXs, business micros and desk-top terminals are indicated by the fact that six prominent cases of recent job losses through closure or contraction (A, G, J, N and two others not referred to above) were firms operating in the more competitive markets. And it is in these areas that, as with consumer electronics, the indigenous computer and telecommunications firms have been weakened by foreign competition; at least in the latter case net job losses may occur too.

Employment–output ratios and employment change

Differences in occupational profiles and activities within the plants largely explained variations in employment–output relationships. Jobless growth or declining employment with rising output was the norm in manufacturing – including in subsidiaries of the same firms located elsewhere in the country. The only exceptions were either where output growth rates were extremely high or where production involved highly customized or project work which allowed no scope for automation or economies of scale. The clearest examples of jobless growth were in semiconductors manufacturing, where increased automation plus larger wafers allowed employment–output ratios to fall. In systems producers, reduced component count and automatic insertion and testing had allowed jobless growth; similar effects were produced by automation in printed circuit board firms. Apart from administration, front-end activities tended to be 'people intensive' and less open to routinization and hence automation. They therefore enjoyed job-creating growth, with the ratio approaching 1:1 in a few cases. One of the contributory factors in this tendency was the so-called software bottleneck. Insofar as the spatial division of labour within electronics is hierarchically organized, it follows that areas like the M4 corridor are less likely to experience jobless growth than others. Where jobs have been lost in the corridor it appears to have been as much a result of competitive failure as of labour-displacing technology or work-intensification (e.g. A, G, K, N). In a few cases, job losses (N) or gains (A) resulted from the redistribution of activities between branches of multiplant firms as technologies, markets and internal organization changed.

Local inter-firm linkages and spin-offs

Subcontracting plays its customary roles: as 'cyclical' or 'capacity' subcontracting, where, to cope with fluctuating demand, firms subcontract out jobs which they might be capable of doing in-house in slack periods but which are not sufficiently sustained to warrant increased capacity, or, alternatively, as 'specialization' subcontracting, to do jobs for which the firm lacks appropriate skills and equipment.[17] Software houses play both roles, whether for unsophisticated clients – mostly users – or for sophisticated clients such as those in our survey. In the latter case software houses are often used for their specialist knowledge of particular computer applications, such as insurance, which many computer firms lack, particularly new foreign entrants.

Though large firms such as O provide most of the software they sell in-house, the small proportion which they subcontract out is quite large in relation to the size of software houses. In other industries subcontracting often gives firms the advantage of access to cheaper labour than they employ in-house but such is the scarcity of software skills that the rates of pay in software houses are frequently higher and their position vis-à-vis large firms stronger than usual. Other examples of subcontract firms are producers of custom hybrid and printed circuit boards, operating in niches defined by specifications, volume and speed of turnaround. For example, a small printed circuit board manufacturer in Bracknell survived by offering quick turnaround on very small batches which larger companies would not bother with.

As one would expect, foreign firms, with their longer planning horizons, lower upstream transaction costs, larger in-house production capabilities and stronger domestic industries, made less recourse to subcontracting and local purchasing than the British firms did. Few of the large firms cited access to subcontractors as a major attraction of the area – though the concentration of software firms in Berkshire, west London and the other home counties was said to be useful by some. Many of the British firms also imported heavily from abroad and sometimes subcontracted work to foreign-owned firms, reflecting the relative technological backwardness of the British industry and the need to keep up with the convergence of telocommunications and computing in IT and the trend towards selling solutions.

Given the dominance of front-end activities in Berkshire and its associated social composition, the concentration of electronics firms is favourable to spin-offs of personnel and technology into new firms.[18] The area provides two of the main necessary conditions for new firm formation – significant numbers of skilled technical and managerial staff who have access to sufficient income and credit to start up on their own and who have inside knowledge of the industry, its technology and markets. Indeed, given the technical nature of the industry, prior experience of electronics, whether in existing firms, government research establishments or universities, is essential. Of these sources, spin-offs of personnel from established firms were by far the most important in the surveyed firms. From the point of view

of the firms losing such workers, it can be seen as one of the disadvantages of working in an industry where small but significant numbers of
employees gain autonomy over their work through having technical
and market knowledge. While spin-offs from government research
establishments may have occurred in the early history of the corridor
there was little evidence of recent examples; indeed given the gulf
between defence and civilian production one would not expect them.[19]
Senker et al. (1985) found in their study of inter-firm relationships
in the Bracknell area that access to local purchasers was important
mainly at the beginning of new firms' lives but that, not surprisingly,
they were not an overwhelmingly important market for more established firms. Localization of spin-offs is due to other factors besides
the existence of a spatially restricted market, such as the tendency
of new entrepreneurs to avoid moving house so as not to increase
their economic uncertainty. Finally, spin-offs may also occur when
established plants *close* and top personnel use their experience and
networks of contracts to set up on their own. Senker et al. discovered a
case of six small firms in the Bracknell area which were set up by staff
made redundant when a large printed circut board plant closed.[20]

Growth and 'overheating' in the M4 corridor

What of the overall impact of the industry in the corridor? Obviously
it is impossible to distinguish with any precision the impact of
electronics from that of other activities like insurance and financial
services and military bases. However, it is even clearer than in South
Wales that industrial development is not an unqualified good for all
but is highly selective in its impact on different groups of people.

Berkshire has had the lion's share of the industry in the corridor
and it is only here that it seems reasonable to speak of an
electronics agglomeration. Yet it is quite different both qualitatively
and quantitively from leading American agglomerations like Silicon
Valley and Orange County in California, each with over 200,000
electronics employees. While it shares with these centres a heavy
dependence upon military expenditure, it does not compare with their
market and technological power. This, and the related backwardness
of the British industry and market, accounts for the fact that while
there have been spin-offs in the corridor their success has been limited. At the same time, Berkshire includes much less manufacturing
in electronics – particularly in the recent foreign entrants – than is
found in the American agglomerations. This probably accounts for
the fact that while local inter-firm linkages are much more common
than in South Wales (thanks largely to the prevalence of less

routinized activities in the corridor) they do not compare with the dense networks of linkages characteristic of Silicon Valley and Orange County (Scott and Angel, 1986; Scott, 1986b; Glasmeier, 1986).

Generally, the electronics agglomeration emerging in Berkshire appears to cohere more on the basis of its local labour market for top technical and managerial skills than on inter-firm linkages. Although these are, of course, also an effect of localization they have reinforced the area's attractions for further development. Outside Berkshire, the impact of electronics is limited; apart from the established aerospace industry it plays a minor role in Bristol's economy and, while Swindon has been successful in attracting prestigious firms, the resulting employment levels in electronics have still not recovered from the major losses in the seventies from two large local British employers. What success it has had owes much to its ability, as an 'expanded town', to grow at a time when development in most other south-east towns was restricted.

Throughout the corridor, but especially in Berkshire, it is the higher-paid and higher-status workers who have benefitted while the situation of the traditional working class has either been unaffected or worsened, not only in terms of job prospects but also in housing. Nor is this bias towards high-status employment limited to electronics. In services, in which Berkshire's employment grew at twice the national average between 1971 and 1981, much of the increase was in the managerial ranks; indeed, the number of manual jobs in services actually fell in the county. How this has happened can be explained by reference to the 'overheating' of Berkshire's local economy.[21] In this respect it is again tempting to draw comparisons with Silicon Valley and its 'urban contradictions' (Saxenian, 1984). There, rising land prices and labour poaching have driven out some electronics firms and the lack of low-cost housing in the area has obliged employers either to bus in manufacturing workers over long distances or set up elsewhere. Activities such as mass production which do not need local skills and linkages have figured prominently in such moves (Scott and Angel, 1986). Yet Berkshire not only has nothing like the number and quality of Silicon Valley's top-skilled scientific workers, it also has no equivalent of its electronics manufacturing plants with their abundant supplies of cheap female and Hispanic labour.

Nevertheless, Berkshire has experienced some problems of industrial agglomeration. Although one of the attractions of the corridor to firms decentralizing from London was cheap land, prices for industrial and office property have risen to some of the highest in the country – £1m and £1.5m per acre in Bracknell and Slough respectively. As firms have grown, many now need room to expand

and this is in short supply as planning restrictions have limited new industrial and commercial development.[22] Though difficult to prove, it seems likely that the overheating of the local economy has actually deterred some inward investors and as much was admitted by some companies which located further west along the corridor.

House prices have risen fast despite considerable new house-building, and this exacerbates the difficulties experienced by firms in recruiting employees, particularly those below professional status.[23] For technicians, with salaries in the £6,000-£7,000 range in the first half of the eighties, the picture is very different. Meanwhile, local buyers have tended to be pushed into second place by new entrants. In 1983–4, 44 per cent of dwellings were sold to people moving into Berkshire for employment purposes, and of these a third were connected with electronics and financial services and a staggering two-thirds of those employed had professional or managerial jobs (Royal County of Berkshire, 1986; Barlow, 1987).

The characteristics of local labour markets are also a mixed blessing for firms, it being easy to recruit but hard to retain labour; depending on the firm's point of view, Berkshire is a good or bad area for poaching. (As we saw in chapter 9, one large military supplier had moved a major division to South Wales largely to escape poaching.) Generally the balance of the poaching is in favour of new firms as against old, and against small firms, unless they are successful innovators in which case they are often attractive to top personnel of large firms who feel they have become too embroiled in administration and distant from the cutting edge of the industry. It is also common for key engineers to be lured away from large employers to work freelance or for specialist contract companies. In those towns to which large offices had decentralized, competition for clerical and secretarial staff was said to be high, though such comments should be set in the context of the unemployment levels of the recession.

Bracknell and Slough in Berkshire and Swindon in Wiltshire provide good examples of these processes. A new town until 1982, Bracknell's proximity to London and Heathrow plus the promise of the M4 motorway attracted ICL, Honeywell, Ferranti, Sperry (now part of British Aerospace) and Racal, with the result that it is now the largest centre of electronics employment in the corridor. During the early days of the new town, firms were attracted by the guarantees of good housing, cheap premises and support for start-ups, all close to London and Heathrow. Now, since the loss of new town status and with the continued build-up of employment, rising land and labour costs have driven some low-tech firms out of the town, and those firms which have moved in in recent years have tended to be ones providing only high-status jobs. Slough, as we noted earlier, is quite unlike the rest of Berkshire in being a thoroughly working-class town; its proportions of households headed by skilled and partly skilled manual workers (29.2 per cent and 19.2 per cent respectively) are significantly above the national average (Barlow and Savage, 1986). However, it is also very different from manufacturing centres elsewhere in Britain and more typical of Berkshire in terms of its relative prosperity; for example, in 1983, it had an unemployment rate of 7.4 per cent – slightly below the rate for the county as a whole – against a national rate of 13.1 per cent, and its rate of owner occupation was 57 per cent, higher than that of Bracknell. But if Slough owes its growth to the boom years of the thirties, fifties and sixties, it is worth noting the contrasts between their social consequences and those of present developments in electronics in the corridor. The former period provided relatively well-paid jobs for skilled manual workers and large numbers of semi-skilled jobs. As we saw in chapter 8, many of these workers originated from the 'distressed areas'. Even though demand for skilled manual workers is still strong in the area, high house prices limit the extent to which it can be met. Generally, however, the contrast with recent times could hardly be greater, for it is precisely the traditional industrial working class which is left out by the development of the modern electronics industry in Berkshire.

The effect of these kinds of overheating has been to push development westwards along the corridor. Swindon is a major beneficiary of this process and is an interesting case of a town which has changed its industrial base and moved up-market. This has resulted in adjustment problems and signs of confused identity. Until recently, there was a local shortage of technicians, though this is being remedied by cooperation between leading firms and the local technical college. The town lacks up-market shopping, entertainment and hotel facilities and so the new executives and professionals have to travel as far afield as Oxford, Bristol, Bath or Cheltenham to find such amenities. While its status as an expanded town has meant that it has been able to provide more housing than many other towns in southern England, it lacks executive housing, though such 'needs' are compensated by the possibilities of living in the Cotswold Hills, with their quintessential 'chocolate-box-top' villages. As always, these are hardly problems that trouble the vast majority of the population. Some will inevitably claim that there might be indirect benefits if they were to be solved (i.e. multiplier effects) but such judgements never take account of the full range of alternative possibilities. Meanwhile, back in Bracknell, where an up-market identity has been achieved, facilities for local working-class people have taken second place.[24]

In short, there has been not only development *in* the corridor but some development *of* it, at least in Berkshire, as a result of the rise of the electronics industry. But insofar as this implies some beneficial local effects, these have been highly selective, overwhelmingly favouring the better-off. Furthermore, and again contrary to the popular view of a haven of free market forces, the corridor's success owes much to state-provided infrastructure and military expenditure, while the most advanced parts of its private industry are foreign-owned.

12

Social Innovations 2
Electronics in the M4 Corridor

Introduction

In chapter 10 we saw how management–labour relations were changing in South Wales as new practices were introduced by inward investors. There, they were applied to activities which are familiar ground to industrial sociologists and students of the labour process – batch manufacturing and mass production. But what of the internal social organization of the plants in the M4 corridor, with their emphasis upon front-end activities? These have received much less attention, particularly from radical social scientists, and yet their significance for both the competitiveness of firms and the character of work is no less than that of the more familiar manufacturing operations in South Wales.

We also saw in South Wales how social innovations within the plants interacted with the characteristics of their localities, particularly regarding behavioural skills and unionization. The same kind of interaction between practices internal to plants and 'outside' in society is evident in the corridor, though with radically different results, as one might expect. In this case the labour market and culture to which new and incoming firms have had to adjust is more permissive than in South Wales. While some industries in the corridor were unionized, they did not have the militant reputation of South Wales, and in any case job losses had greatly reduced their size. From the point of view of electronics firms the corridor is perceived as a 'middle class' society and a 'union desert'. While this is not entirely an accurate impression there is certainly less of a social legacy to be remade by management or, conversely, to be unlearnt by labour, than in Wales. Yet, while the industry has been welcomed by a pro-growth coalition in South Wales, it has received a much more mixed reception in the most successful parts of the corridor.

Perhaps surprisingly, given the different context, some of the same themes that were discovered in South Wales – regarding recruitment, flexibility, employment status, motivation and communication – came through again. Another similarity is the striking contrast between the management–labour relations of the more recent inward investors, in this case largely American, and those of the oldest British plants, with the situation in the newer British plants tending to lie between these two extremes. Again, the picture was far from static, with the more established British-owned plants attempting to imitate some of the American firms' practices in order to improve their competitive position. Yet there are also significant differences from the Welsh situation. These stem from three sources: the different range of activities and mix of occupations (more front-end, less routinized etc.); the differences in the kinds of companies (mainly British and North American, but no Japanese companies of any note); and lastly the differences in the industrial history, social composition and character of the labour markets in the localities.

In this chapter we shall look into these matters and make comparisons with the findings of chapter 10. We shall begin with an outline of some of the key types of skills, workers and labour processes, detail the social innovations in the corridor, and then evaluate the attractions of the area for the firms, ending with a brief discussion of local reactions to growth.

The workforce

Just how different the electronics industry in the corridor is from that of South Wales was evident in the comparison of their skill profiles (see figure 8.6 on page 148). In Berkshire – the most developed part of the corridor – the proportions of managers and unskilled manuals were 13 per cent and 12 per cent respectively, compared with 3 per cent and 66 per cent in South Wales. Given the growth of employment in front-end activities and the shortages of skilled workers such as software engineers, these figures are probably now underestimated; in Berkshire, for example, it has been estimated that no less than one in three job vacancies are in electronics (Royal County of Berkshire,

1985). For some professional jobs, firms estimated that there were three vacancies for every two qualified applicants. Experienced software engineers and marketing staff, the latter a result of the increasing commercial orientation of much of the industry, were the most commonly cited areas of shortage. In both cases firms were seeking a combination of behavioural and technical skills. Of course, all skill shortages derive from a lack of training, but lower down the scale, e.g. technicians, an additional reason for shortages is the high price of housing in much of the corridor. As we shall see, this has had profound effects on the industry, producing a 'poaching war' in the Thames Valley. Despite the existence of informal agreements between major employers on local salaries, there has been a bidding up of salaries and complaints were made about the growth of a pool of overpaid but often mediocre engineers.

As regards gender, in contrast with the situation in South Wales where 58 per cent of the electronics workforce are women, the proportion is only 29 per cent in the corridor (1981 ER II data). This reflects the much smaller proportion of the operator jobs in the corridor. Firm N was typical: at its M4 plant – which involved mainly front-end activities – the male–female ratio was 70:30, while its Scottish manufacturing plant had a ratio of 30:70.

Of course, the concept of masculinity involved in managerial and technical work has nothing to do with physical strength and hard manual labour since none is needed in the industry; rather, masculinity both refers to – and is confirmed by – the competent exercise (by men) of power over technology and other people. Indeed, for men, these qualities compensate for the absence of any opportunity for the display of physical toughness in the workplace.

Sometimes the taken-for-granted associations of particular jobs with specific genders were upset and confused by technical change.[1] The most obvious case is the arrival of QWERTY keyboards in workstations and personal computers (PCs) on the desks of male employees. (While one company did claim to expect all new recruits to have typing skills, we suspect that few men use their PCs for typing.) One manager of an American firm claimed that managers were now typing when dealing with electronic mail, while secretaries were preparing data for reports on PCs, but such 'role reversals' hardly alter women's position.

The dominant view of the role of women in the front-end parts of the industry was probably represented by a glossy pamphlet produced by ICL, called 'Women in ICL'. This contained examples of women workers, most of whom had risen to higher positions than is usual and who were clearly in the 'superwoman' mould, capable of managing both a development team and a family with two small children. In other words, no concessions were made to women's unequal place in society and the apparent message to women readers was 'these women have managed to cope with the dual role, why not you?'

One kind of concession which has received wider attention is the employment of skilled female home workers or 'pregnant programmers' as one firm called them. Another firm, 'F International', is based almost entirely on such workers. These women usually start off in the normal way, working in offices, and the decision to let them work from home probably has more to do with the desire of their employers to retain skilled labour than to some new kind of gender enlightenment. However, this situation is the exception rather than the rule.

Compared with South Wales, a much larger proportion of workers were in a position of strength vis-à-vis their employers on account of their specialist technical and market knowledge; indeed they might be thought of as a new labour aristocracy, albeit one noted for its *lack* of, and indeed contempt for, unionization. This position gives them a leverage far beyond that possessed by those who have only less skilled and more dispensable labour power to sell.[2] Also work was more open ended, given the dependence on product and market development, project or very small batch production rather than repetitive production. So the scope for routinization, standardization and deskilling of work was strictly limited and success of the plants depended in no small measure upon managers and workers using their own initiative and continually tackling work which was at least to some degree novel. While we saw in South Wales that the most advanced firms had realized the advantages of giving workers more discretion, the typical M4 plant, with its emphasis upon front-end activities and its large proportion of highly skilled workers, has no option but to do so. At the extreme, knowledge of strategic importance to a company – e.g. regarding new products – might exist largely in the heads of key workers. Such individuals naturally have a strong position, not only in the labour market but sometimes in being able to set up in business on their own. Even though the majority of front-end employees have less power than this, their position is still appreciably stronger than that of most workers in the South Wales plants.

These characteristics set up familiar tensions in management–labour relations. Insofar as the work is more challenging and less routinized, workers are more likely to find it fulfilling. It is therefore not unusual to find 'professionals' working late without extra pay, though often this is expected and rewarded indirectly. For example, particularly in some of the more research-oriented plants, researchers were not bound by a fixed working day but would still would still work 'ridiculous hours', and in a couple of cases offices were kept open round the clock.[3] While such workers might seem to require less motivation, their autonomy and the nature of the work also gives them more scope to 'coast' without this being detected by their superiors. This of course applies to managers themselves too. Even where they do not attempt to do this there are better and worse ways of reinforcing individual motivation and of translating it into commercial results, and firms which lack these organizational skills are likely to be uncompetitive. Projects are not executed, nor are products designed, marketed and manufactured, by individuals but by teams, and their success depends on their cohesion. In other words, despite the reliance upon highly skilled individuals, organization is still crucial: 'whizz-kids' are not enough.

Software projects are a good example, often requiring many person-years of work by a team. As a non-routine form of production, a new technical division of labour for the project may have to be worked out in advance and, if necessary, adapted as unforeseen problems arise.[4] Given the research or problem-solving element of such work, not all needs can be anticipated: to a certain extent their discovery is part of the work to be done. In the nature of software, coordination between different elements has to be extremely rigorous, which means that exchange of information between project members has to be intensive; indeed a high proportion of errors stem from ineffective communication between team members. As the size of project teams grows, the number of possible interactions between members increases exponentially, and so perennial attempts have been made in the industry to modularize and structure the work. An early response to these problems was to Taylorize it, increasing specialization and deskilling jobs wherever possible. This appears to have been largely a failure, chiefly because of the resulting restriction of learning-by-doing and the exacerbation of the problems of communication between workers. Instead of a split opening up between more conceptual work (systems analysis) and execution (programming), in many cases a new multiskilled analyst-programmer grade has emerged, indicating convergence and broadening of skills (Friedman and Cornford, 1985). If, as we suggested, the limitations of Taylorism are becoming apparent in more routine kinds of work, their unsuitability for highly skilled non-routine work is all the clearer. More successful have been 'software tools' for aiding productivity and the technique of structured programming, which provides a logical, hierarchical approach to programming, and one which has been not only pushed by management but embraced by programmers wanting to make their job easier.[5] Contrary to early labour process theory, not all changes in work are imposed on reluctant workers. Nevertheless, the problems of formalizing the 'black art' of software design and of organizing projects so as to produce accurate code quickly and within deadlines and budgets remain; it is widely believed in the industry that individual productivity can vary by up to a factor of ten. This 'people problem' is central to the phenomenon known in the industry as the 'software bottleneck'.

Theorists on the Left have sometimes seen these problems as consequences of the class power of programmers and the resulting failure of management to take command of the knowledge used in the work process. In other words, the bottleneck is seen as resulting from a modern version of traditional craft workers' power to coast under the noses of less knowledgeable managers (see CSE Microelectronics Group, 1980). If this is so then the power is surely exercised individually rather than collectively in most cases, for computer programmers are hardly known for their radicalism. We are more inclined to believe that the bottleneck arises for an additional and simpler reason (though one which seems inconceivable to the Braverman school), namely the intrinsic difficulty of software design.

As in South Wales, whether conducive to competitiveness or not, the various elements of management–labour relations frequently operated in a mutually reinforcing way, so that each element had multiple rationales and effects.

The vanguard firms

The vanguard position in management–labour relations in the corridor was dominated by seven large North American firms, all of which have chiefly front-end activities in the corridor. At the other end of the spectrum were several established British companies, with the remaining more recent plants tending to lie between the two extremes.

Recruitment

Recruitment procedures were extremely rigorous, with multiple interviews ranging from at least two for semi-skilled production and clerical staff to as many as *eight* for professional staff in one big-league computer firm.[6] For non-professional staff, the emphasis was often placed not on prior technical skill but on outlook and adaptability – on willingness to train, retrain and rotate on a regular basis if need be. As for professional R&D and marketing staff, formal technical qualifications were less important than actual technical experience, especially when combined with the social skills needed for effective communication with colleagues and clients. Rigorous recruitment procedures were now more than ever essential because it was becoming standard practice to manage growth with as little addition to the regular 'core' workforce as possible. Consequently, the use of contract staff, especially in the data-processing field, and 'temps', for clerical work, seems to be a major trend in the computer industry.

Central to the management–labour relations of the vanguard firms were *flexibility* and *motivation*. Whereas in back-end activities flexibility tended to mean willingness and ability to switch between tasks which were already largely predefined by management, in front-end activities it meant flexibility in the more open-ended situation of project, small batch work and the development of new markets and products. Indeed, given the need to react quickly to changing customer needs, flexible response was crucial. Accordingly, job description were kept vague, as in the leading firms in Wales, only in this case professional employees were frequently expected to be responsible for building up their own work pattern in the process of developing the company's business. (This suggests that once again control need not be a zero-sum game between management and labour, and the leading new firms managed to combine increased managerial control with a greater degree of responsible autonomy than that allowed to workers in the old plants.) Flexibility, along with productivity and coordination, were further facilitated by shifting staff between different activities within the firms, by single status regarding conditions of work, individualized payment systems and employment security (see below).

As we have already noted, whatever the intrinsic interest of some of the work, it does not necessarily follow that skilled workers will use their relative autonomy to their employer's advantage, and the most skilled always have the option of taking their skills elsewhere. Consequently, the leading firms went to extraordinary lengths to retain and motivate such people and to regulate their work in ways conducive to high productivity. Professionals often became aggrieved if their job remained unchanged for more than two years and so these firms made extensive use of job transfers and *internal labour markets* to motivate them. Flexibility, motivation and use of internal labour markets are therefore mutually reinforcing. However, unless employment growth was rapid, this strategy ran into the problem of 'career headroom': not everyone can occupy top positions simultaneously and it can be counter-productive to create excessive optimism about promotion. To ease this difficulty, the leading firms created elaborate but well-defined and accessible career paths, involving horizontal as well as vertical movements, the latter being largely restricted to professional staff. The emphasis on internal labour markets was also useful as a way of retaining those already attuned to the 'corporate culture' and hence reducing the costs of training in behavioural skills.

Management strategy

Because individual motivation need not translate into the effective *social* organization required for competitive operation, considerable efforts were made to monitor, counsel and guide front-end employees, both individually and in teams. All the leading firms used some form of individual review procedure. This is a system in which each employee meets his or her superior on a one-to-one basis at regular intervals to review progress and jointly set objectives, which would then be compared with actual performance at the next meeting and appropriate adjustments made. This was usually done annually or quarterly, though in one firm all employees except the most lowly – in other words, including senior management – were expected to write monthly reports on progress. Another firm had, in addition, what it termed a system of 'pastoral care' in which salaries were reviewed four times a year and career progress once a year. In most cases there was also an open-door policy enabling dissatisfied employees to go over their supervisor's head. This policy has the added benefit for the firm of making recourse to union intermediaries unnecessary and workers often found they got a quicker response to complaints than they would have done through official union channels. So while most managers in the M4 plants (and many of those in South Wales) had an individualistic conception of industrial relations, according to which all problems pertained purely to individuals and hence were to be resolved at that level, the ideology was most successfully put into practice in the leading M4 firms.

Pay structures

Pay structures provide a further motivating force, not just in terms of relatively high average rates but in being based on individual merit. While a small band of firms in South Wales found it necessary, if difficult, to legitimize the uncoupling of pay from the going rate for the job, most firms in the corridor had accomplished this without any great difficulty, the main exceptions being in some of the manufacturing operations. In the leading firms pay increases were purely merit-based and there was no collective bargaining. Increases ranged from 0 per cent to 8 per cent and, in one case,

16 per cent though naturally they were made from within a fixed budget, so there was a zero-sum element in the competition between individuals. It is therefore not unusual to find two people working on the same job whose pay differs by as much as 50 per cent (hence the opposition when an older, unionized firm introduced them – see below). In most cases pay was reviewed annually, though in one case there were quarterly reviews. While individual performance assessments obviously feed into management's decisions on pay, for obvious reasons most tried to conceal the link between them by excluding discussions of pay from review meetings. Only in some of the smaller research-intensive firms was employee stockholding favoured as a further form of financial incentive, presumably because it was in these firms that the risks of losing key staff were greatest.

The most striking feature of management practice within the leading firms was the immense amount of effort invested in directly communicating with employees, whatever their grade. Management in the new entrants in South Wales may have been close to and knowledgeable of the labour process, but here managers were often *directly* engaged in the more strategic labour processes, e.g. team leaders on R&D projects. The significance of this communication in front-end activities such as sales is reflected in the very high manager–subordinate ratios – often 1:8 or even 1:6. Given the greater autonomy of many of the professional jobs, the starkly delineated authority relations between management and labour in the shopfloor arena – in both Wales and the corridor – are decidedly blurred at this professional level. Great importance is attached to regular *face-to-face* contact between managers and workers, a bilateral process unlike the depersonalized and unilateral provision of 'mere information' via noticeboards and the like. This strategy, in one firm practised regardless of grade or seniority, represents something more than a cosmetic public relations facade: it helps to sustain high employee motivation and low labour turnover and it serves as a very important means of surveillance.

Horizontal lines of communication between staff were also given considerable attention and most favoured open-plan offices, even for the most senior staff, both for this reason and for closer surveillance. Behavioural skills were prioritized not only for jobs involving contact with customers but for facilitating internal communication and hence efficiency. For example, good communication between marketing, applications and sales is crucial for winning the custom of large, sophisticated and discriminating customers. New employees might be mystified by managers' concern with their communication skills, perhaps because these are frequently considered to be unchangeable personal characteristics, but the rationale in terms of collective productivity is sound. Indeed, this emphasis given to effective internal communications in the leading firms is also consistent with the findings of the Sappho project on innovation (Science Policy Research Unit, 1972), which showed that the quality of internal and external communications differentiated successful from unsuccessful innovators more clearly than any other variable.

In one of the leading firms, staff would have to give regular presentations on their work in small groups for evaluation and suggestions. At these sessions the presenters would be expected to include not only highlights of their work but 'lowlights', and in order to discourage defensive attitudes and concealment of problems the firm ran courses in 'constructive confrontation' for staff. Essentially these meetings were an equivalent of quality circles in manufacturing – a forum in which technical and organizational problems could be discussed and resolved collectively. In another firm, small work teams were set objectives each year and then given discretion on how to meet them.[7] At the same time as 'responsible autonomy' and discretion are encouraged, management also tries to monitor and raise individual productivity. In some cases this is done by harnessing peer-group pressure and individual competition; for example in software projects it is becoming common for team members to have to check each other's work. Most of the leading firms used similar methods and all attached great importance to getting people to *reveal* rather than *bury* problems.[8] However, one of the problems of individual performance reviews is that managers are loath to make too many criticisms of their teams in case it reflects badly on themselves. At any rate, methods such as these amount to a determined effort to create a *continual institutional learning process* as a way of improving competitiveness.

Taken together, such practices add up to distinctive 'corporate cultures' and a radically different experience of work from that in traditional firms, so much so, in fact, that new entrants frequently underwent a period of 'culture shock' for much of their first year. While growth is a precondition for the successful development of these management–labour relations, in the case of two semiconductor firms they proved effective enough to allow them to get away with a '125 per cent working week' (with no extra pay) in one and short-time working in the other. In the latter case, the majority of staff continued to work without pay on the days off.

In other words, while the leading firms stressed *individualized* relations for the purposes of motivation, surveillance and control, they had to stress *teamwork* for ensuring that work was done effectively. This of course reflects a universal feature of capitalist labour processes but these firms had more success than most in ensuring that the collective forms of organization essential for production did not become the basis of defensive collective action by labour, although they were helped in this respect by contingent factors such as the small size of teams and the limited duration of most projects. (In the two vanguard firms which had staff associations, however, management regarded them warily lest they formed a basis for collective action.)

The belief that 'small is manageable' appeared to be even stronger here than in South Wales – though no doubt the technical characteristics of the work, in particular the absence of mass production, allowed it to be realized more fully. Large firms would try to avoid what they saw as the inevitable rigidities and communication problems of large units by breaking their operations up into small divisions. In other words, many of the firms were organized to maintain *socially optimal* economies of scale, with activities grouped on product or market, rather than hierarchical, lines so that units were fairly self-contained and had self-imposed employment thresholds. The perceived merits of this strategy are that greater intimacy facilitates managerial control while at the same time preserving the 'small firm' environment and sense of identity much favoured by professional staff. Even within these units work was usually organized on the basis of small teams.

There was also a problem that senior staff are becoming frustrated by their growing distance from the cutting edge as promotion within large organizations led to increasing involvement in administration. One firm employing 500 people deliberately split its operations into three units to avoid further losses of such people to small start-ups. Views varied on just how small units should be – from a maximum of 400 in a computer firm to 25 in a firm manufacturing chip-making equipment whose workers were spread out over no less than nine sites!

These, then, are the kind of management–labour relations found in the leading firms in the corridor, particularly the recent North American entrants. At the other extreme are plants belonging to a handful of large established British firms involved in producing computers, telecommunications and military equipment.

The traditional firms

In contrast with the vanguard firms, these had more bureaucratic forms of organization and control of labour, little job rotation, less flexible working and greater status differences within the workforce, weaker vertical communication in terms of control over individual workers and weaker lateral communication between different workers.[9] The older firms also lacked the rapid growth which was an enabling factor, and in part a consequence, of the management–labour practices in the American firms. Some were quite heavily unionized and pay was mostly negotiated collectively with pay scales relating to seniority and particular jobs.

In one old firm there were particular difficulties with bureaucratic and social divides between marketing and development. Communication between these – vital for selling solutions – was limited by overt hostility between the two groups, while software work was hampered by the tendency of programmers to work individually on 'their' programs without coordinating with others. Forms of individual motivation were weaker; as one manager who had moved from a traditional British to a US entrant put it: 'in the old job people didn't say they worked there, they said they merely attended'. And, whereas the 'corporate culture' was discussed incessantly in his new job, the term would not have been understood in his old one. Given the nature of much of the work, the managerial strategy of responsible autonomy was more fully developed in the new firms in the corridor than in those in South Wales. However, just because highly individualized forms of vertical communication and motivation were used less in the old British firms, it does not follow that they exhibit greater solidarity and cooperation among workers at the same level; if anything, we suspect the contrary.[10]

As regards use of internal labour markets versus external recruitment, the older British firms were caught in a dilemma: while they were increasingly aware of the value of internal labour markets for motivation, they lacked the employment growth to allow career headroom and in any case they also felt the need to attract 'new blood' from outside. On top of this, they were more vulnerable to poaching of their best staff by new firms and software houses, and we suspect that, while their more ambitious staff may be tempted away, others might be discouraged by the alien corporate cultures of the vanguard firms.

As in Wales, most firms operated a core–periphery division in their workforces, only in this case the periphery included not only clerical temps but highly skilled workers, particularly software engineers working freelance or on contract from software houses. In some of the larger plants, these numbered about 30. While in a few cases, firms might not have been able to find full-time work for such people, in large part this reflects the labour market strength of such workers, insofar as many prefer the higher pay, greater autonomy and more varied work of contract employment.

Again, as we saw in chapter 10, it was evident that intensified competition had encouraged an *imitation effect* between the old firms and the new, again often in a rather limited, half-hearted manner which reduced the effect of managerial innovations and increased scepticism and cynicism towards them. In chapter 10 we noted the paradox of trade unions in a traditional labourist area in some cases facilitating new work practices. Here, unions and labourist attitudes to industrial relations are much weaker, yet what little influence the unions had in the old plants in the corridor was not used in this facilitative manner. On the other hand they have not proved a serious barrier to the introduction of new work practices. Far more important as an impediment is the sheer inertia – at managerial levels as well as in the office and on the shopfloor – of bureaucratic organization and entrenched traditional work practices, albeit those of mostly white collar workers. (Academic readers should not be surprised by this; the same would surely be true of universities and polytechnics!)

Increased competitive pressures in military equipment markets were 'wreaking havoc' in one firm, obliging it to change its management systems to try and increase productivity and eliminate the slack periods which had formerly characterized its project work. Some of these projects lasted for 15 years and not surprisingly this was hardly conducive to high productivity for the remoteness of the deadlines allowed bureaucracy, boredom and complacency, or 'cobwebs', to develop in the teams. While the introduction of more competitive forms of contracting and costing has obliged such firms to introduce tougher supervision, their situation is still different from that of the more commercially oriented firms in which projects tend to be measured in months rather than years, and pressure on employees is more intense.

Some of the older firms had added a non-negotiable individual merit element to their traditional collectively bargained payment system, though the size of these payments was smaller than in the vanguard firms, most of which had no collective pay awards. Not surprisingly, in the most unionized firm objections on the grounds of divisiveness and inequity (the 'blue-eyed boys') were raised, but the changes went through nevertheless, though the management would have liked to have had a purely merit-based payment system.[11] Quality circles and team briefing have been tried out in the same firm, 'flexibility' has become a buzz-word, multiskilling, youth and behavioural skills have been emphasized in recruitment, and attempts are being made to assess the productivity of sales and development personnel for the first time.

As so often happens in weaker firms, senior management finds itself in the contradictory position of trying to win more active consent while upsetting old work patterns and cutting the workforce – in firm A, to the tune of 10,000 jobs throughout the UK. Similar pressures are being felt in the newly privatized British Telecom, where privatization has been accompanied by a sudden emphasis on behavioural and communication skills and the renaming of programmers as 'managers'. Though the latter change is purely cosmetic, the former is directed at raising software productivity by reducing errors arising from communication problems within teams.

Intermediate cases

Just over half the firms visited occupied an intermediate position in this spectrum between the large recent North American entrants and the old British firms, most of them perhaps being closer to the former than the latter. They were often smaller than the plants at the two extremes and this in itself made for flexibility and minimal bureaucracy. While most had less formalized corporate cultures and systems of motivation and individual assessment, some of the more research-intensive firms used employee stockholding as a motivational device. In the new primarily manufacturing plants, there had been some reduction in demarcation and one firm had introduced individual assessment and merit payments in addition to collective awards, but generally, aside from the question of employee resistance, the nature of the work gave less scope for such practices.

Unionization

Trade unions were conspicuously absent in the majority of plants, the few exceptions being some of the older established plants or more recent manufacturing branches of large, unionized British companies. Most managers saw unions as an unnecessary tier between them and their employees and many said – often using the same words – that they would see it as a sign of failure if there were ever calls for unionization. Yet the feeling that unions could not offer employees anything that they had not already got was not limited to managers but was common among non-manufacturing staff.

There are several reasons for this resistance to unionization. Firstly, the predominance of professional employees, who are less inclined to

be collectively organized than are manual workers, enables individualistic norms to be prescribed for *all* workers. Secondly, skill shortages
are such that many professionals can win high salaries by individual
pay bargaining – market forces are in their favour and they stand to
gain little by joining a union as things stand at present. Thirdly, in
some cases a sizeable proportion of the workforce worked on their
own outside the plant, e.g. field service engineers, and therefore had
little opportunity, let alone desire, to organize collectively. These
features reinforce the fourth factor, namely the unorganized labour
traditions in the area which were partly responsible for the firms'
choice of this location in the first place. Fifthly, while large plants
account for a considerable proportion of employment, small firms
account for 66 per cent of all electronics firms in Berkshire, and the
intimate social relations in such firms constitute a formidable deterrent to trade union organization; indeed one union official considered
this dispersed nature of the workforce to be the main problem from
an organizational point of view. Sixthly, it seems far too simplistic to
suggest – as is frequently done – that anti-union corporate strategies
have been successfully imposed on apathetic or recalcitrant workers.
This reading fails to appreciate that the rationale behind pastoral care
is not limited to the negative goal of forestalling unions; it is, rather,
the more ambitious goal of winning over workers of all grades so that
they *actively* identify their interests with those of the company. To
the extent that this is achieved – and a high growth context provides
fertile conditions for a positive-sum game – trade union affiliation
often appears unnecessary and anachronistic. Lastly, the pervasive
influence of military work in the corridor cannot be discounted as a
further barrier to trade union organization. If the effect is anything
like that in the USA, military procurement will be instrumental in
the formation of 'conservative and pro-military cultural attitudes'
(Markusen, 1985). Whatever the precise combination of factors, the
central point is that non-unionization has been sustained in large part
because there appears to be little or no *demand* for union recognition.

Naturally, the anachronistic, 'dirt under the fingernails' image of
trade unions is one which many managements are happy to project,
but some unions have indeed failed to appreciate the nature of the
kinds of work and workers at the front end of the industry. In one
plant carrying out development work on military products, a union
recruitment meeting was held at which the employees turned up in
their usual suits and ties only to be confronted by a leading trade
unionist wearing a T-shirt bearing a revolutionary slogan and delivering revolutionary rhetoric having no perceptible relevance to their
situation. Needless to say, the meeting was a failure. Later, however,

another leading trade unionist from a different union recruited a few people by adopting an entirely different style – both in dress and rhetoric – in which he appealed to their sense of status as professionals needing the support of an organization that understood their needs.

Several unions have launched major recruitment drives, mostly without success. For example, Technical, Administrative and Scientific Staff (TASS) held a meeting in Bracknell – the 'capital' of the M4 industry – on skill shortages, but hardly any non-union workers attended. Given the vested interest of skilled workers of remaining in short supply, this is hardly surprising. The EEPTU has a full-time officer trying to get union deals in Berkshire's electronics firms, though with little success; for example, while it has several agreements with firms such as Ferranti and 3M in different parts of the country it has not been able to get one in their Bracknell plants. However, the polarization of employment between manuals and clericals, whose labour market position is extremely weak, and highly skilled workers, who are scarce, is producing a polarization in pay, thus creating resentment in the lower grades and the possibility that the latter may become more susceptible to recruitment.

Even if we discount the penchant for fads and gimmicks in management circles, there is no doubt that the new management–labour practices have real economic effects in raising productivity and competitiveness. Even so, these new practices are still difficult to introduce into traditional plants, even in a 'permissive' area such as the corridor. Equally, the experience of working in the newer, un-unionized plants seemed to be better than in the old plants, both objectively in terms of security, pay and conditions, and subjectively in terms of motivation. We should perhaps repeat that the development of these superior conditions in the new firms has nothing to do with altruism but is a management strategy for enhancing profit, albeit one influenced by the market power of highly skilled professionals. Acknowledging this does not mean that the advantages for labour are illusory. Such conclusions, of course, are unpalatable to the Left, but they expose the error of assuming that management–labour relations are simply a zero-sum game, in which capital's gain is inevitably labour's loss. Nor should we assume that every characteristic of management–labour practices is simply reducible to an expression of the capitalist character of the firms. Professionals do not work extra hours simply because capital forces them to do so, nor because they are falsely conscious. A combination of forces encourages such workers to work hard: not so much the fear of unemployment, as such a prospect is less likely for professionals, and not only management pressure, but interest and craft pride in the job (for example, most programmers feel bad if their program fails to work). This craft pride can of course be present even where the product is a socially damaging one, such as a missile system. And, after all, if workers in less skilled jobs usually give a measure of active consent (Burawoy, 1979), it is hardly surprising that professional workers should do so. To some extent, similar attitudes could and surely should be expected in non-capitalist organizations too (and management would be needed no less!).

As in other respects, in the information technology (IT) industry the lead in the development of new forms of industrial relations appears to have come from IBM.[12] In IBM generally, each employee has an annual interview with his or her superior for 'appraisal and counselling' and setting objectives. Status is unobtrusive, training has traditionally been far more extensive than in rival firms, and moving people between different jobs is used as a way of transferring technology and improving coordination. IBM's plant at Greenock in Scotland has long been organized in units of ten employees, each with a leader who not only supervises them but determines the relative pay of each individual by putting them in one of five grades (Jay, 1972).

Finally, to return to our general theme of the interaction of industry and locality, although the corridor is so different to South Wales, our conclusions are similar. Thus, while it cannot be overemphasized that the significance of place, in particular the permissive work culture and image of the corridor, is seen as a crucial enabling factor for good management–labour relations in electronics firms, this social context is itself malleable within limits.

Location decisions and evaluations

We have seen how the location decisions of firms coming to the corridor have also been affected by the characteristics of its labour markets and work culture. However, other, non-labour, considerations have been influential to a greater extent than in South Wales. These chiefly concern access to customers – not only for marketing but for maintenance, consultation and support. Proximity to government research establishments as customers was important at first but is now only significant for a few firms specializing in scientific markets. Arms firms also still find it useful to be located within two hours' drive of London and major government research establishments and military bases. Now, as commercial uses of electronics have expanded and the internationalization of production and consumption has increased, access to customers at home and abroad has become more important; indeed, this was the most widely noted non-labour-related requirement. As we noted earlier, the IT market is heavily polarized on London, with its company headquarters, public sector administration and financial services. All the foreign firms attached great importance to Heathrow Airport, primarily as a gateway to and for foreign customers and secondarily for ease of access within and from the UK for their own staff.

Some readers may find it strange that an industry whose products include advanced communications equipment should attach so much importance to face-to-face contact both within their organizations and with customers. But communication through such technology cannot rival the richness (and greater pleasurability) of face-to-face contact, with its ease of trust-building and scope for serendipity and 'duty-free' perks!

The need for access to suitably skilled labour was almost universal and the corridor, especially Berkshire, was repeatedly cited as the best area in Britain for electronics skills. This was even the case where the numbers of skilled personnel needed were small, for they were nevertheless seen as critical for success. Also, with the increased importance of software relative to hardware and the associated shift from manual craft and semi-skilled labour to non-manual and 'professional' labour, the latter's influence upon location has grown stronger. Yet there is more to this tendency than a simple locational effect, for behind it lies the 'free rider' syndrome common amongst firms, according to which they try to benefit from the training provided by others, rather than incur costs in doing the training themselves.[13] The localization of skills in the corridor is both cause and consequence of this process. Not to acknowledge this syndrome is to overlook the national lack of training and to be complicit in the myth that skills are innate or merely self-taught.

Nevertheless, this still leaves the question of why highly skilled labour, apparently so mobile, should be so localized. Countless surveys of electronics engineers find that such workers rank location – meaning environment and schools – higher than salary, and most of the firms in our survey confirmed the importance of the residential qualities of the area. Even firms which can afford to recruit over a much wider area have to be able to retain their recruits. They therefore have to weigh the advantages of choosing an area attractive to such people against the high risks of poaching: normally the former wins and hence local labour markets *are* important to them.[14]

Readers who are geographers will probably recognize the notion of 'locational preferences' in this argument about the localization of front-end activities (Keeble, 1976). While this seems relevant, it is so only with three crucial qualifications. First, the preferences are cultural products of specific groups in society, not innate preferences of all members of society. As Massey puts it, this is

> a process of social cumulative causation: the very fact that people of the same social standing live round about is confirmation in itself of arrival at a particular status, and confirmation too of a set of values reinforcing them and ensuring their reproduction. And the jobs follow. (Massey, 1984, p. 142)

Second, they are only realizable by those with scarce skills and high incomes. Third, they are only realizable within areas in which it is feasible for employers in the industry to locate; for example, prior to the construction of the M4, preferences for Newbury's desirable environment were much less realizable than those for places such as Maidenhead or Bracknell.

However, the companies themselves have another reason for preferring sunbelt areas such as the corridor to older industrial areas. When asked why they had not chosen to go to a development area (for some of the manufacturing operations, inaccessibility would not have been a problem) several mentioned the expense and trouble of traditional and ununionized workforces as a problem and the union-free environment of the corridor as an advantage. Company A had concentrated most of its software work in the corridor rather than at its provincial sites and it was the corridor which was most favoured in its recent reorganization. According to the firm, this geographical shift was an unintended consequence of reorganization and differential rates of growth but unions saw it as a deliberate policy to concentrate activities where unionization was weakest.

Finally, it must be said that there was also a barely concealed 'herd instinct' in the choice of location; Berkshire was popular because 'all the major electronics companies in the world were there'. Moreover, one sometimes actually finds rival foreign firms locating equivalent operations in the same areas (e.g. European headquarters in Geneva, semiconductor design centres in Bedford, British headquarters in Swindon). To be located within a 'prestige environment' was itself important for many firms, from the point of view of both attracting top skills and impressing customers, and this influenced their choice of site and premises. One American firm admitted that it had initially made the mistake of locating on Merseyside where it had difficulty in recruiting engineers and had subsequently moved south to solve this problem and to confirm its own high-status self-image. In Bristol, some firms chose the Aztec West high-tech part mainly for reasons of prestige, despite complaints about the practicalities of the site. A few other firms had found that they could more easily retain staff and gain prestige in the eyes of customers by taking over large country houses. In general, as the 'mythologies of Maidenhead' suggest, the corridor is ideal for sustaining elite lifestyles. Only a few parts of the corridor are unsuitable, and one of these, 'proletarian' Slough, offsets its social disadvantages by its excellent access to Heathrow.

Berkshire's anti-growth coalition

In South Wales we saw how business, labour and regional state institutions formed a pro-growth coalition. While the corridor is certainly not lacking in state expenditure, albeit in forms other than regional policy aid, most of the corridor lacks local employment policies to attract firms. Swindon is the major exception, with its high-profile advertising campaigns, but it has also gained from overheating and the lack of incentives to incoming firms in neighbouring Berkshire. Here the County Council is not allowed to solicit new entrants and indeed is considered positively unwelcoming by the industry.[15] Nevertheless, the growth of electronics employment in Berkshire between 1971 and 1981 was the highest in Britain. More remarkable is the fact that the county not only lacks a policy to attract firms but now possesses a powerful and vocal *anti-growth* coalition – a spatial coalition centred on fears of congestion, pollution, erosion of the green belt and deteriorating services and quality of life. Animated by affluent grass-roots pressure, Tory MPs from the sunbelt are up in arms against their own government in what amounts to an unholy alliance with the more depressed regions of the country, because

> We want to put a brake on development. Concern to protect the environment has now become a political fact of life. Even the truest of blues has turned Green ... Areas just beyond the boundaries of the belt, such as Berkshire and North-East Hampshire, feel themselves under threat. A successful regional policy is necessary if a growing national imbalance is ever to be corrected. Prosperity must be returned to the Midlands and the North. Britain cannot be permitted to live and work within 30 miles of Heathrow. (Critchley, 1987)

However, there have been some pro-growth elements. In the past, there was Reading's fostering of office building prior to 1974 and Bracknell's new town status prior to 1982. At present, as one would expect, major firms wanting to expand locally are naturally pro-growth, not only in electronics but in services such as retailing. But while employment growth is generally welcomed, the only local groups who welcome the associated increases in housing are farmers and landowners wanting to realize the potential value of their land. Apart from these the main pro-housing groups are non-local: the House-Builders Federation (HBF) and central government. On present growth rates, Berkshire is set to become the most densely populated non-metropolitan county in the south-east (Barlow and Savage, 1985). It is in the Home Counties that the HBF have pushed hardest for the release of land and house-building rates in Berkshire have been high at a time when they have stagnated nationally. Yet house prices continue to rise above the average for the south-east, not only because of the level of demand but because the major builders now concentrate heavily on expensive executive housing.

The anti-growth coalition includes some strange bedfellows. First there is the conservation lobby. Much to the embarrassment of Tory politicians in central and local government, this is dominated by Conservative voters, many of whom paid inflated prices to live in a rural area, only to find it becoming less rural as subsequent development has been allowed. A second and highly ambivalent group is formed by established manufacturing businesses needing mainly non-professional labour which find rising house prices, land prices and wage costs a problem, and sometimes move out as a result of this. Additional housing suitable for workers would ease two of these problems, but attracting more professionals into the area, for example by allowing more electronics firms in, would worsen them. (The Council would actually have liked to attract traditional manufacturing activities but this is forbidden too.)

Similarly ambivalent is the third group, the representatives of labour, who would favour more blue collar jobs and low-cost housing but realize that future employment growth, house-building and house prices are likely to continue to favour professionals at the expense of the traditional working class. They are therefore against more employment and housing growth unless it radically changes character.

Despite the fact that the local Tory politicians are strongly Thatcherist, they have reluctantly had to take on board conservationist policies, at least regarding housing. Some major development initiatives, such as a large high-tech business park between Bracknell and Wokingham, have been refused. However, the coalition has also had its losses, most spectacularly when central government intervened to force the county to build an extra 8,000 houses (now popularly known as 'Heseltown' after Michael Heseltine, the minister who overruled the structure plan), though even then conservationists managed to reduce the number going to Bracknell. Presumably, behind Heseltine's intervention lay the belief that an obstruction to growth here might retard *national* economic growth. In other words, the area is having growth forced upon it in the 'national interest', despite broad-based opposition, just as public spending has been cut back in the metropolitan areas on the same grounds. The difference is that while the latter strategy entails confrontation with the government's enemies, the former antagonizes many of its supporters.

This is particularly ironic in the Berkshire case because the area is seen as a valuable ideological resource, an exemplar to the country at large of 'growth without aid'; hence the refrain that 'Britain needs more Berkshires' (*The Economist*, 1982). Although this is of course nonsense, because it ignores Berkshire's place within wider spatial divisions of labour and takes no account of 'hidden' state expenditure, this has not prevented it from gaining wide currency.

13

'Sunrise' Industry, Innovation and the State
The Neoliberal Regime

The uneven fortunes of localities and regions should not obscure the larger canvas, where the relative decline of Britain's national economy, or at least the industrial component, is the main issue. Indeed, by post-war international standards Britain as a whole might be considered to be something of a 'regional' problem. For example, with the exception of the south-east, every UK region suffers from higher unemployment and lower output than the EEC average (Commission of the European Communities, 1984a). On another front it appears that large sections of UK industry have retreated from higher value-added products and within the 'high research-intensive sectors' themselves there has been a significant erosion of the UK's position (Pavitt, 1980; Midland Bank, 1986). Then, in 1983, there was the 'final ignominy', namely the UK's first deficit on its manufacturing trade since the industrial revolution (Gardiner and Rothwell, 1985). It is against this sombre background that current and alternative political strategies have to be assessed.

Whatever their differences, all political parties seem to agree that the electronics revolution is a historic opportunity for Britain to reverse decades of relative industrial decline, provided that a comparative advantage can be created in the supply and use of electronics technology. On both sides of the political spectrum the state is deemed to be crucial to this project, albeit in different ways: on the Right the main task of the state is to create a 'climate for enterprise'; on the Left the state is expected to play a much more *dirigiste* role. Behind these superficial distinctions lie very different conceptions about the adequacy of the market as the chief mechanism for promoting industrial innovation on the one hand and for meeting social needs on the other.

However, the Left's long-standing theoretical debate on the capitalist state leaves much to be desired in terms of defining what role the state might play in a socialist strategy for innovation. This is no accident. With few exceptions the Left has been curiously reluctant to address itself to concrete levels of analysis, preferring to dwell on the abstract character of the capitalist state and on the ways in which the state supposedly delivers the functional conditions for capitalist accumulation.[1] But, as we saw in part II, there are great differences in the way that actual capitalist states play their roles, with varying degrees of success. In the more functionalist Marxist conceptions these concrete permutations tend to be filtered out, as if they were incidental to the essence of the capitalist state. In what follows we take the view that the concrete permutations matter a great deal because some states have been far more functional than others, and this helps to explain the changing global pattern of uneven development, especially so far as the electronics industry is concerned.

Our aim in this and the following chapter is to look at the prospects for reversing attrition in Britain, taking account of the specific character of British capital and the British state. In this chapter we present a brief outline of the neoliberal strategy and go on to assess neoliberal policies for the electronics industry and for the wider economy. Before doing so, however, it is worth noting the role which the British state has played in skewing the resource base of the economy in such a way as to make it more difficult for Britain to achieve a viable civilian economy.

Defending the realm – depleting the (civilian) economy?

A central component of Britain's relative industrial decline has been its poor record of technological innovation (Freeman, 1979). This might seem paradoxical given the enormous resources which the state has committed to R&D over the years. But this paradox begins to disappear when we inspect the nature of Britain's public and private R&D efforts. Superficially it might seem that Britain's record here has been comparable with that of other major Organization for Economic Cooperation and Development (OECD) countries inasmuch as its total R&D expenditure, expressed as a percentage of gross domestic product (GDP), seems to hold up well. However, GDP *per capita* is small in the UK and, when the resources devoted to R&D are adjusted to take account of relative population size,

Britain drops to the bottom of the league table. Secondly, a higher proportion of Britain's GDP is devoted to defence R&D than in any other OECD country except the USA. When defence R&D expenditure is excluded Britain ranks below all its major competitors.

The political priorities of the British state are positively perverse here: over 50 per cent of total government R&D expenditure is now devoted to defence, while defence activities account for only 6 per cent of GDP and a mere 3 per cent of total exports. Moreover, of total government R&D expenditure within industry – as opposed to that within government – 81 per cent went into defence in 1980. In contrast, 'other industry, trade and employment' received only 4 per cent; hence this category, which is the main source of funds for the civilian engineering industry, represents a pitifully small fraction of total expenditure. Perhaps this low political commitment to civil R&D would not matter so much if industry-financed R&D was as robust as in other major OECD countries. Unfortunately, Britain's record here has been dramatically worse than its rivals over the past two decades. Between 1967 and 1982 industry-financed R&D increased by only 0.9 per cent per annum in Britain, compared with 9.8 per cent in Japan and 5.9 per cent in France and West Germany (House of Lords, 1983; Marsh, 1985). Surveying this dismal picture the Department of Trade and Industry (DTI) was forced to conclude that a 'combination of different indicators all point to the declining industrial impact of the UK's civil R&D effort' (Department of Trade and Industry, 1985, p.4).

Defending the realm is an enormously costly business: in 1986–7 the total Ministry of Defence (MoD) budget amounted to £18.5b, of which £8.2b was devoted to equipment, a budget which had increased by some 40 per cent in real terms since 1979 (Ministry of Defence, 1986a). To these direct costs one must add the substantial, if unquantifiable, opportunity costs; that is, the potential civilian benefits that are foregone as a result of such a massive commitment of human and financial resources to a sphere which has few commercial spin-offs (House of Lords, 1986). While the relative decline of the UK manufacturing sector is a familiar enough theme, the fact that this sector has become increasingly dependent on defence is perhaps less well known. For example, in the decade to 1984 defence production increased its share of manufacturing GDP from 6.3 to 12.3 per cent, while its share of engineering GDP increased from 17.3 to over 30 per cent in the same period (Kaldor et al., 1986). What we have here is a growing militarization of the UK's sunrise industries, since aerospace and electronics between them account for over 50 per cent of the defence equipment budget. Indeed, this military bias is even greater on the R&D front, where the MoD accounts for over 80 per cent of all government R&D funds to the electronics industry (Cabinet Office, 1986). Furthermore, at a time when the civilian economy is desperately short of a skilled workforce, it is estimated that some 25–30 per cent of all qualified scientists and engineers are now 'colonized' in defence activities (House of Lords, 1986; Kaldor *et al.*, 1986). Small wonder that the UK's major electronics firms have devoted their leading-edge activities to the defence sector when this sector has been so privileged by successive governments.

The low commercial spin-off from defence appears to be part of a wider problem, which is that Britain has poorly developed mechanisms for diffusing the results of state-sponsored R&D programmes in and beyond the defence sector. As a result Britain can be classified as a 'mission-oriented' country, in the sense that it has been primarily concerned to promote projects with an accent on defence or national prestige, such as Concorde. In contrast we can speak of 'diffusion-oriented' countries, such as Sweden, Japan and West Germany, where the emphasis has been placed on upgrading the technological capacity of all sectors. In Britain's case

> mission-oriented research has tended to yield few direct benefits, while possibly crowding out a substantial share of commercial R&D. The indirect spin-offs have been low, creating a "sheltered workshop" type of economy: a small number of more or less directly subsidized high technology firms, heavily dependent on and oriented to public procurement; and a traditional sector which draws little benefit from the high overall level of expenditure on R&D. (Ergas, 1986, p. 20)

This state of affairs is partly attributable to a pervasive lack of incentives and penalties in Britain's mission-oriented R&D programmes (particularly in defence and telecommunications) and to the fact that public procurement agencies tended to neglect export potential when specifying their equipment needs. But it is also the case that potential benefits have been denied to the wider economy because public resources have been trapped in the sheltered divisions of a small circle of large firms (House of Lords, 1983; Ergas, 1986).

In a later section we argue that Britain is ill-equipped to capture the benefits of a robust diffusion policy because of a number of weaknesses in its user sectors. In this section we have simply tried to show the extent to which the state has skewed the resource base of the economy towards military activities and how, in terms of R&D support, the civilian engineering sector has a Cinderella look about it. Even in those civilian sectors where the state has had a powerful procurement role, such as telecommunications, the direct and indirect benefits have fallen far short of what might reasonably have been expected. With the exception of GEC and similar companies, the main beneficiaries of Britain's mission-oriented regime, the British state has been far from functional in the wider project of regenerating the civilian industrial sector. This has made it that much easier for an 'anti-state' programme to emerge.

Re-asserting the market: the neoliberal strategy

Changing the rules of engagement

Reversing decline through the market was the avowed aim of the Thatcher Government when it assumed power in 1979. At the forefront of its neoliberal programme were a number of interrelated objectives: (a) to combat inflation, (b) to cut government expenditure and roll back the public sector, (c) to chasten trade unions, and (d) to deregulate the economy. Behind this bold agenda lay an unmistakably behaviouralist interpretation of Britain's post-war industrial decline. In the neoliberal scenario successive governments had failed to confront organized labour, deemed to be the chief source of inflation, and this, coupled with an overregulated economy, had rendered the market less of a social force. Consequently, trade unions and a resource-hungry state were identified as the two main obstacles to industrial recovery.

The Thatcher regime heralds a decisive break with the past in at least two ways: in the overriding priority accorded to the fight against inflation, and in the attempt to absolve the state of its traditional responsibilities. As regards the first break, the government calculated that a resolute anti-inflationary strategy could yield dividends on a number of fronts. For instance, sweating inflation out of the economy would require high interest rates and a strong exchange rate and these, in turn, would force industry to become 'leaner and fitter', as it was put. The aim here was to promote a bout of 'creative destruction' in the hope that liquidated assets would be assimilated by more productive competitors (Burton, 1979). Such an exacting environment thus held the promise of lower inflation, a more innovative industrial sector and, because of the higher unemployment entailed by this strategy, a chastened trade union movement.

The second break consists of an attempt to absolve the state of responsibility for any economic objective other than inflation. Strictly speaking, the only targets which the Thatcher Government set for itself were for the money supply and government borrowing. The refusal to set goals for employment and output stemmed from a desire to shift the main responsibility here onto the shoulders of capital and labour, stressing the effect which their actions had on the level of economic activity. This signalled new rules of the game: since 1944 it had been accepted that the 'duty' of government was to pay homage, if nothing else, to such targets as high employment and growth etc. (Buiter and Miller, 1983).

If the duties of government are conceived in traditional Keynesian terms then the neoliberal strategy can indeed be seen as 'non-interventionist'. But this is far too simplistic. It is not that the Thatcher Government is averse to intervening, but that the targets and methods of intervention are no longer those of traditional Keynesianism. Under neoliberalism the duties of government are largely to combat inflation and create a climate for enterprise and here Thatcherism has been vigorously interventionist – witness its determined interventions to maintain high interest rates so as to reduce inflation regardless of the costs on the industrial economy; its privatization of more than £11b of state assets; and its legislative offensive against trade union activities.

By and large the neoliberal emphasis has been upon indirect rather than direct forms of industrial intervention, one index of which is that the DTI budget was cut by some 65 per cent between 1979 and 1987. At the outset practically all forms of direct industrial aid were thought to be inconsistent with a 'social market' economy. Since 1979, however, the Thatcher Government has been forced to abandon this pristine position and to admit that direct aid for information technology (IT) was unavoidable. This reversal came at the end of 1980, prompted by a belated recognition that overseas support schemes threatened to disadvantage UK industry. As a result a series of new support schemes emerged, mainly under the banner of the Support for Innovation programme, to promote the development and application of IT. But not even these schemes for innovation have escaped stringent economy measures. By late 1984 all this aid was frozen, ostensibly because available resources had been exhausted by a higher than expected demand. When the freeze was eventually lifted the Support for Innovation budget had been reduced by £10m to £298m, with little or no increase planned for the future, and stricter conditions were applied to existing aid. Furthermore, there was also a shift in the balance of future aid, away from direct industrial aid to awareness campaigns (Large, 1985). Spending on 'enabling' technologies has increased while aid to mature industries (such as coal, steel and vehicles) and regional policy aid have been cut sharply. This is part of a new emphasis designed to shift direct aid away from sectors to enabling technologies (Department of Trade and Industry, 1983c). This was an explicit admission that much of UK industry was still unable or unwilling to utilize advanced IT, a point to which we return later.

Overall though, the role of the DTI has become increasingly marginalized under the Thatcher Government. Cutting the Support for Innovation budget was justified on the grounds that corporate profitability had improved, and therefore firms should carry more of the costs of innovation. Although it is officially the main department of state for industrial policy it is worth noting that its total budget of £1.6b for 1986–7 is now nearly £7b less than the equipment budget of the MoD!

Despite the grudging creation of new DTI support schemes for innovation the Thatcher Government still perceives these as subordinate adjuncts to the macroeconomic strategy:

> Of overriding importance is the objective of controlling inflation ... this will make a more substantial and lasting contribution to the improvement of industry's international competitiveness than fine tuning by the instruments of industrial policy and the major means to that end is the determined control of monetary growth and of public expenditure. (Department of Trade and Industry, 1983a)

Just as this anti-inflationary strategy is presented as the major form of 'industrial support', large claims are also made on behalf of deregulation and inward investment. In the neoliberal repertoire this triad constitutes the recipe for recovery, rather than direct intervention via the DTI. But the triad embodies threats as well as opportunities, especially for those sections of capital which are not up to international competitive standards.

Within Western Europe the UK has gone furthest in deregulating its economy. Although this has allowed vast amounts of capital to be exported from the country it has also led to an influx of inward investment. One of the proudest boasts of the Thatcher Government is that under its tenure the UK has become one of the most attractive European locations for US and Japanese capital. There are a number of reasons for this, as we saw in chapter 8. Here it is worth noting that Britain now has a corporate tax structure second to none in terms of its liberality. And, as a tax haven, Britain is now on a par with the Cayman Islands (Economist Intelligence Unit, 1985). However, the two factors which Thatcherism itself uses to sell the UK abroad are a pro-business government and 'non-unionised or single union facilities' (*Business Week*, 1985a).[2] While direct job creation is traditionally seen as the main rationale for inward investment, this is not the overriding issue for the Thatcher Government. Listing the benefits of inward investment the Invest in Britain Bureau claims that foreign firms bring 'new technology, new management styles and attitudes, the injection of capital investment, the generation of exports and new jobs' (Invest in Britain Bureau, 1983, p. 2). Of these the government appears to lay greatest stress on 'Japanese-style' management practices (Department of Trade and Industry, 1983a). Overseas firms, especially those from Japan and the USA, are seen as potential 'tutors' to both labour and management in Britain; they are portrayed as vehicles for the re-industrialization of Britain and therefore they are 'treated in the same way as British owned companies' (Invest in Britain Bureau, 1983, p. 2). In other words, the Thatcher Government hopes to regenerate the UK economy on the basis of a cosmopolitan capitalism, rather than one in which domestic capital is prioritized. Indeed, no government has gone so far in obliterating the distinction between British-owned and British-based capital. What matters in this scenario is not so much the nationality of capital but its competitive capability.

So, apart from a limited amount of direct industrial support, the main thrust of the neoliberal strategy is to create a climate in which private capital can assert itself. However, if this strategy has had debilitating consequences for organized labour as a result of mass unemployment and restrictive union laws, it has not been entirely functional for industrial capital either. Even in sunrise sectors such as electronics, many captains of industry are far from happy with neoliberalism.

Neoliberalism and the electronics industry

The condition of the electronics industry in Britain has become something of a controversial issue. On one hand the Thatcher Government claims that 'there has been a shift in recent years towards greater optimism about its future' (Department of Trade and Industry, 1984a). The National Economic Development Committee (NEDC), on the other hand, perceives the industry to be in 'relative decline' for the reasons referred to in chapter 8.[3] Although part of the official optimism can be discounted as a propaganda exercise, there is also a more substantive difference here. The NEDC appears to be more concerned about the plight of the indigenous industry while the government, as we have seen, includes both indigenous and foreign capital in its definition of the 'British' electronics industry. For example, the warning of an 'impending crisis' in the IT sector refers to the danger that, on present trends, the UK 'will not have an independent broad-based IT industry by the end of the decade' (Information Technology EDC, 1984, p. 9). In this view it is essential to preserve and promote a viable indigenous industry, otherwise the UK economy will increasingly have to depend upon imported technology and systems. And, should this occur,

> We shall lag behind the leaders in applying these by two to three years. The systems will not be as well adapted to our particular

needs. Originators of the systems will be more strongly placed as a result of their prior knowledge and expertise to compete in a wide range of related business. Exploitation of the technology and of systems employing it may be constrained by a foreign company or government. (Information Technology EDC, 1984, p. 5)

Many of these problems can be categorized as the negative effects of external control, an issue generally considered only in the regional context. But these problems are now surfacing at the national level too. What needs to be remembered here is that external control is not seen as a problem at all in the neoliberal scenario; on the contrary, inward investment is seen as a wholly positive development. Apart from this issue the main problems facing the indigenous electronics industry have changed little from those identified some years ago by the NEDC. The chief ones are the following:

1 UK firms have a poor record of commercializing technology, or of translating inventions into commercial innovations; this reflects a failure to integrate R&D with production and marketing;
2 there has been a widespread failure to develop world market strategies, and consequently UK firms have been denied the benefits of volume production;
3 part of the explanation for point 2 is that the leading firms have been overly dependent on public sector markets, and state dependence acted as a surrogate for overseas markets;
4 most firms suffer from short-term corporate horizons, one of the main reasons for this being the need to 'satisfy a financial community anxiously watching each quarter's figures';
5 insufficient resources are being devoted to R&D for civilian products, and this partly stems from an 'overcommitment' to military work, from which there is poor commercial spin-off;
6 many firms suffer from an acute national shortage of IT-related skills, and this technical skill shortage is compounded by the fact that social relations within the firms are not conducive to rapid technological change, since management is often tied to restrictive protocols when communicating with its workforce;
7 finally, there has been a comparatively weak interface between government and industry, with the result that a 'commonality of purpose' is less evident in the UK relative to the UK's overseas competitors (Electronics EDC, 1982; Information Technology EDC, 1984).

Radical initiatives were necessary, claimed the NEDC, if these problems were to be resolved. To this end it recommended a collaborative effort between government, industry and the trade unions to identify a medium-term strategy for the electronics industry. In this context it proposed a package of measures, including a new government–industry interface for planning and development, a more active and better organized public procurement policy, geared to indigenous suppliers where new products were concerned, enhanced support for civil R&D, better provision of equity and loan finance over the long term, a national training initiative and a more discriminating approach to inward investment.

This strategy, with its corporatist overtones, was anathema to the Thatcher Government. Rejecting anything that smacked of a tri-partite strategy, the Industry Secretary argued for a 'hands-off' approach, claiming that the private sector alone should decide the shape, size and direction of the electronics industry, because it alone knew where the 'prospects are brightest'. Apparently unaware of the inconsistency, the Industry Secretary nevertheless castigated UK firms for using defence contracts 'as a cushion rather than as a springboard for risky international business outside the defence field' (Morgan, 1983b).

However, this hands-off posture raises some pertinent questions. First, it rests on the strange claim that the state has little or no legitimate right to play an active role in the electronics industry – even though the state is itself the largest single customer and a major source of R&D finance for the industry. Second, by ceding full authority to the private sector, it allows narrow corporate interests to decide the 'brightest prospects' in an industry which has enormous implications for the entire economy. The case of semiconductors is instructive here. Because the private sector decided that its interests were best served by withdrawing from mainline microchips so as to concentrate on specialized chips for the protected military market, Britain was left without a major indigenous source of supply. The creation of Inmos, the state-backed microchip company set up by the last Labour Government (but privatized under Thatcherism) was an admission that corporate interest was not at all synonymous with the public interest. These two questions – lack of accountability and the efficacy of the market – are not issues at all for the Thatcher Government, for it is profoundly undemocratic with respect to industrial power, while the efficacy of the market is assumed to be above dispute. However, the question of political leverage can be a real problem in a hands-off strategy particularly if, as the above criticism implies, firms are not responding to the neoliberal design or if foreign governments are not similarly committed to a hands-off approach. As we shall see, these problems have forced the government to temper some of its pro-competitive spirits.

There are three main ways in which the Thatcher Government has sought to stimulate the electronics industry, apart from those mentioned in the previous section; these are through deregulation, collaboration and inward investment. The first has had the greatest impact in the defence and telecommunications sectors and, since this was intended to be so, we shall focus on these here. Deregulation, or liberalization, is partly designed to overcome a long-standing problem, namely the fact that the leading firms are too heavily dependent on public sector markets. In the defence sector the MoD is imposing a more liberal procurement policy in an effort to get more 'value for money' for itself and the wider economy, the main elements being the use of more competitive contracts, fixed price contracts, greater pressure on contractors' profit margins and a growing readiness to buy from abroad. Furthermore, after seven years of feast under the Thatcher Government, the defence budget will have declined by 7 per cent in real terms between 1986–7 and 1988–9 (Bloom, 1986). Though still an enormously important market for electronics equipment, around £2.5b in 1986–7, the leading firms are now having to adjust to a chillier climate. In fact, GEC, Racal, Thorn-EMI and Ferranti collectively declared over 2,000 job losses in their defence divisions in 1987, partly as a result of this new climate.

But it is in the telecommunications sector where the government has wrought the greatest change. By deregulating the market, privatizing British Telecom (BT) and licensing a rival carrier, in the form of Mercury, the government has turned the UK into the most liberal market outside the USA. In many ways this sector was a ripe candidate for radical political surgery: the export record of the equipment industry had been dire and the development of System X, the UK's digital electronic exchange, was way behind schedule, with the result that it has yet to win a major export order. Much of the blame for this pedestrian performance lay in the stultifying, club-like relationship between the Post Office and its 'big three' suppliers, GEC, Plessey and STC, who were content to service their captive domestic market. Well aware that all previous attempts to reform this club had failed, the Thatcher Government decided to destroy it (Morgan, 1987).

This radical new departure had two broad aims: to create a more competitive market and to promote a stronger industry. In practice these aims have pulled the government in different, and sometimes contradictory, directions. The best illustration of this was the conflict between liberalizing the market and privatizing BT. Anxious to strengthen BT so as to obtain the best possible share price, the government eschewed the counsel of its more pious supporters who, logically enough, felt that BT ought to be broken up if a more competitive market was the name of the game. Furthermore, in allowing BT to acquire Mitel, one of its foreign suppliers, the government was judged to have seriously impaired competition in the UK market (Monopolies and Mergers Commission, 1986). Despite the formation of Mercury, BT's dominance in services and equipment is such that a near-private monopoly has supplanted a public monopoly. (These criticisms in no way imply, however, that we should return to the 'good old days' when BT was a captive market for its domestic suppliers – see chapter 14.)

The impact of these policies has been mixed in other ways too. Large business users are the major beneficiaries: they have more freedom as to where and what they purchase while their tariffs have been reduced, in contrast with residential tariffs which have increased (Hills, 1986). The major losers are the indigenous equipment suppliers: gone is their once captive market, and GEC and Plessey now have to compete on equal terms with Thorn-Ericsson (a 51 per cent British-owned firm based on Swedish technology) to supply BT with digital electronic exchanges. Although the government hoped that its 'cold shower' policies would goad the likes of GEC into becoming more competitive, it has been badly disappointed with the response so far. As for BT, which is now subject to a light regulatory reign, it has lost no time in pursuing an aggressive commercial strategy. It now feels less bound to procure from its traditional UK suppliers and it is shedding labour rapidly: having shed some 20,000 jobs in the four years to 1986, it intends to dispose of a further 24,000 jobs between 1986 and 1990. Given the growing disjunction between the UK telecommunications market and the indigenous telecommunications industry, it is no surprise that the liberalization of this market corresponds to a growing trade deficit in telecommunications equipment, so much so that the UK is now the major deficit country in the EEC (European Research Associates, 1986). There is also the question of BT's R&D facilities. Once an enormous, if under-used, public resource, these have been transformed into a proprietary asset, geared to BT's private market goals. Whereas France and West Germany are able to use their public telecommunications authorities as vehicles for IT-related strategies in conjunction with indigenous suppliers, BT is no longer in a position to play such a wider role. In contrast, the UK's path is being driven by the narrow commercial interests of BT and the large business users on whom it depends for so much of its profit. In the USA, the model for the Thatcher Government, these large business users are now busily constructing private telecommunications networks which enable them to 'by-pass' the main network and this, in turn, leads to higher charges for residential and smaller business users.

In telecommunications, a sector of great importance in itself, and as an end-user of software and microchips, the Thatcher Government has tried to create a 'Little America'. But while the USA deregulated in a context where its domestic producers were strong, the UK has done so without having this advantage. Hence the major UK suppliers feel deeply aggrieved by the fact that their government has liberalized at home without securing reciprocal arrangements abroad (Morgan, 1987).

While deregulation is perfectly at one with the tenets of neo-liberalism, collaboration, the second main policy thrust, is more of a departure forced on government by the growth of overseas support programmes. In 1981 Japan frightened its competitors when it announced a state-sponsored initiative to develop 'intelligent' or fifth-generation computer systems. In view of this threat the Thatcher Government was persuaded of the necessity for a collaborative programme of pre-competitive research in the enabling technologies of IT (i.e. microelectronics, software engineering, artificial intelligence and so-called man–machine interfaces) (Department of Industry, 1982a. The result, the Alvey programme, is a five-year effort involving government, industry and academia, costing around £350m in all. Of this, £200m is government funded, a sum equal to total government expenditure on microelectronics in the five years to 1983 (Arnold and Guy, 1986). What concerns the government, however, is that precedents like this tend to generate a demand for follow-on programmes. With the Alvey programme due to expire in 1988, the government is now faced with a demand for an additional £300m to support another IT programme, on the grounds that 'a laissez-faire approach will not suffice' in IT (IT 86 Committee, 1986).[4] Here it is worth noting that although the government is itself the largest single user of civilian IT, public procurement has not featured much in this collaborative form of intervention. In fact, the government's own record of investment here is extremely poor in comparison with its counterparts in the USA and France.

In addition to the Alvey initiative there is also an array of supra-national collaborative ventures, like Esprit in IT and Research and Development in Advanced Communications Technology in Europe (RACE) in telecommunications, both sponsored by the EEC. As we saw in part II, some of the main problems with these supra-national efforts stem from political differences between member states. The timing and efficacy of these EEC initiatives have been seriously threatened by the Thatcher Government. Alone among the member states, it refused to accept the level of financial resources deemed necessary by all other EEC countries (Hills, 1986).

The third way in which the government hopes to stimulate the electronics industry is by encouraging multinational 'tutors' to locate in the UK. Although inward investment is promoted in all fields it has been most vigorously championed in electronics. But, as chapter 8 shows, the effects are not all positive. Indeed, inward investment is another issue where the leading indigenous firms feel badly betrayed by their government. This domestic lobby is opposed to what it sees as strong international competitors gaining access to a liberal market and being subsidised, via regional grants etc., to do so. Once here, it argues, these foreign firms exacerbate Britain's already acute skills shortage, displace existing output and employment and make it more difficult for indigenous firms to enter new growth markets (UK Information Technology Organization, 1986).

These criticisms neatly capture some of the main conflicts of interest here, even if they are motivated by special pleading. But what this lobby fails to realize is that the Thatcher Government is fully prepared to accept these costs. As an internal DTI paper puts it, 'a corollary of introducing a competitive spur is acceptance of the risk of displacing some of the output and jobs provided by existing IT companies'. As regards skills, it says 'inward investment can aggravate skill shortages but the key question is whether the skills are being optimally used' (Department of Trade and Industry, 1984b). This reinforces the point that, for Thatcherism, it is not so much the nationality of capital that counts as its competitive capability.

By exposing indigenous firms to the pressure of deregulation on one side and inward investment on the other, the Thatcher Government hopes to shock them into becoming more dynamic national and international competitors. The major indigenous firms have found this to be a bruising experience, so much so that they claim that none of their overseas competitors has been so disadvantaged by such 'unpatriotic' government action (General Electric Company, 1985). Even if these criticisms are treated as 'tactical' differences they are sufficiently important for us to question the claim that 'big capital' has supported Thatcherism because it sees this as the only political force capable of delivering 'capitalist solutions' (Hall, 1985). The fact is that the Thatcher Government has pursued a high-risk strategy in the electronics industry: it has weakened the traditional strongholds of the indigenous firms in defence and telecommunications and, by promoting the UK as a primary site for multinationals, it has made it that much more difficult for the former to diversify into civilian markets. Furthermore, overseas firms tend not to bring their most advanced technologies to Britain and, to the extent that they do not balance their intra-corporate trade, inward investment may not have as much of a positive influence on the UK's balance of trade as is commonly thought. Indeed, if present trends are not reversed, the UK deficit in electrical and electronic goods is forecast to increase from £2.6b in 1984 to £8.3b in 1993 (Cambridge Econometrics, 1985).

There is nothing necessarily chauvinistic about wanting to preserve and promote a viable indigenous electronics industry. The consequences of not doing so will have adverse effects on both suppliers and users, as we saw earlier. As regards the civilian sector, it is not at all fanciful to foresee the UK becoming a 'technological colony' of large offshore companies, with these deciding what, when and how things are designed and produced in the UK (Maddock, 1983). Such a situation contains an obvious threat to organized labour, given the strong association between non-unionism and no-strike deals on the one hand and inward investment on the other. These new-style labour practices are of course being taken up by the indigenous sector too and, as far as the government is concerned, this indicates the potential for British firms to emulate their more successful rivals, especially the Japanese. However, as we shall see, there the similarity ends.

The broader front: towards a neoliberal revival?

One of the conditions for reversing industrial attrition in Britain is the ability to deploy advanced technological capabilities throughout the economy, in the service sector as well as in manufacturing itself, and in so-called sunset, as well as sunrise, industries. Failure in this respect has been one of the more important factors behind the relative decline of British industrial capital throughout the century.

Yet for Thatcherism the two chief sources of Britain's economic problems were trade unions and the public sector. By attacking them the government hoped it had laid the basis for an industrial revival on neoliberal lines. While it has met with considerable success in subduing organized labour, rolling back the public sector and reducing inflation, a broader revival has not been forthcoming. Partly to overcome the problems of a diminishing customer base at home, Britian's major manufacturing companies have fled further abroad in recent years, mainly to the USA and Western Europe. Between 1979 and 1986, for example, the top 40 increased their overseas workforces by 125,000, or 15 per cent, and cut their British workforces by 25 per cent, equivalent to 415,000 jobs (Labour Research, 1987).

Behind its ideological self-assurance the Thatcher Government has gradually come to appreciate the unpalatable fact that many of the problems here have little if anything to do with trade unions or the public sector. For example, a government inquiry into the scope for technological change in the manufacturing sector painted an unflattering picture of management:

> Much of it emerges as impervious to technological change, ignorant about its implications and ill-equipped to deal with it. A quarter of managers surveyed said they had made no significant changes in their production processes in the last five years, and a third said new technologies had no impact on them . . . and 40% of companies said they had no strategy for coping with technology at all. (de Jonquieres, 1985c)

This dismal picture can be found in a whole series of reports, official and unofficial (National Economic Development Committee, 1983a; Snoddy, 1985). But government itself is also culpable here because, as we have seen, its own investment in civilian IT is lower than that of other major governments.

This poor capacity for innovation is generally held to be most acute in the development areas. In recent years the official diagnosis of the regional problem has shifted. For decades the chief weakness of the development areas was attributed to their overdependence on mature industries, but now it is ascribed to a 'low innovation potential'. At bottom this refers to the low R&D capability within both independent firms and branch plants and to the fact that these regions have a poor capacity to generate new firms (Department of Trade and Industry, 1983c). But the neoliberal strategy offers little hope of regenerating these depressed regions. Indeed, if regional policy is any guide, the political priority attached to regional development has been severely devalued under Thatcherism, because regional aid has declined by some 50 per cent in real terms since 1979. This is part of the neoliberal emphasis on regional autarchy, i.e. the depressed regions are expected to rely largely on their own indigenous resources.

Although regional policy has not been completely jettisoned, the government has tried to emphasize the contribution which factors other than regional policy might play in its scenario of regional development. Firstly, trade unions are enjoined to become more tractable with respect to both pay and working practices, so as to enhance the competitive performance of existing industry and to create a flexible social image for potential new inward investment (Joseph, 1981). Secondly, the depressed regions are encouraged to avail themselves of national enterprise schemes which, it is claimed, will enable them to generate indigenously based new firms. The first prescription places the onus for regional development almost entirely on 'natural adjustment' in the labour market, a euphemism for lower pay, the potential of which is discounted even by the Confederation of British Industry (CBI). The second prescription conveniently ignores the structural bases for historically low levels of new firm formation in these depressed regions. Here we refer to the fact that the class structure of these regions, because it is mainly composed of 'lower' social strata, is far less able to generate new firms compared with the English sunbelt. So the fortunes of the depressed development areas depend, now as before, on the fate of the UK economy and the priorities of the British state.

In fact, the changing composition of state expenditure under the Thatcher Government has actually worked against the development areas in favour of the more prosperous areas, as a result of the decline of regional policy and the rapid growth of the MoD's procurement budget, which is heavily biased towards southern England. And, because the take-up of national schemes for new firm formation and advanced technology is also biased in this way, these policies tend to reinforce existing inequalities in innovative potential between regions (Morgan, 1985, 1986).

Returning to the poor innovation record of Britain as a whole, two further major causes require note: the woefully inadequate provisions for ongoing training; and the 'tyranny' of the City. Although weak training provision is a long-established problem it has been exacerbated by the neoliberal strategy: in 1981 the government abolished 17 of the 24 Industry Training Boards, ostensibly because statutory training programmes were ill-attuned to corporate require-ments. However, the trend towards voluntary training initiatives has been a great disappointment to the Manpower Services Commission (MSC), so much so that the MSC now speaks not of a skill shortage but of a 'skill crisis' because, it says, Britain's workforce, *from top management down*, is technically impoverished compared with other major countries. As a result, the MSC has warned that 'if Britain is to travel the technological road to full employment, we must re-skill the workforce, from top management to shop floor, and we must do it now' (Pike, 1986). This skill crisis is most acute in relation to IT professionals, where demand is outstripping supply by a large margin. But, as the MSC indicates, the problem is not simply one of unfilled vacancies; it is also that the general level of skill in a whole range of occupations – in production, maintenance and marketing through to senior management – are conspicuously low by leading international standards (Manpower Services Commission, 1984; National Institute of Economic and Social Research, 1985; National Economic Development Office, 1987). With notable exceptions Britain's managerial class has placed a low premium on reskilling itself and its workforce. As a result it is perhaps not surprising to find British managers treating R&D and training as expendable items, indeed as items of current expenditure rather than items of long-term investment. Such cost-cutting has of course been encouraged by the Thatcher Government's high interest rate regime and by other well-established short-term pressures.

This brings us to the last major obstacle to innovation, namely the lack of 'patient money' from the City (Advisory Council for Applied Research and Development, 1978). Despite reassuring noises from the City, a significant number of British firms think and act in the short term because they feel themselves to be 'under the tyranny of the immediate' (House of Lords, 1985, p. 67). The problem is not shortage of money capital but, rather, the lack of long-term lending facilities: this forces companies to gear themselves towards respectable quarterly or half-yearly profits. The search for short-term payoffs in turn means that the appraisal of advanced technologies too often shows poorer returns than the use of existing technologies, hence discouraging innovation (New, 1986). As one DTI official has argued, 'if UK companies acted like Japanese companies their share price would collapse and they would be vulnerable to takeover' (Gillan, 1986). 'Ironically, it was during 1986 – the so-called Industry Year – that the ranks of British industry felt the 'tyranny of the immediate' as never before, when assets of more than £11b were acquired in the biggest rash of takeovers since 1968.

For a government which is second to none in extolling the judgement of the market, it seems odd that the DTI has joined the ranks of those who castigate the City for failing to take a long-term view of advanced technology, and that the Treasury should be censuring industrialists for their 'short-termism' with respect to R&D and training (*Financial Times*, 1986). It is even more remarkable in view of the Thatcher Government's own refusal to invest in long-term public infrastructure for the sake of short-term financial 'savings'.

Though grateful for the fact that Thatcherism has enhanced managerial power vis-à-vis organized labour, this no longer seems enough to the captains of industry. For example, according to a CBI survey of industrial problems, 46 per cent of managers cited high interest rates as the key problem, 54 per cent cited exchange rate volatility, while 60 per cent said the key problem was the fact that 'the government has written off manufacturing' (Lipsey, 1985). Official equanimity in the face of a declining UK product base, and the government's savage attack on the House of Lords report on overseas trade, which warned of an impending industrial crisis, spell a new phase in the political denouement of industrial capital in Britian. For its part the Thatcher

Government takes refuge in the belief that losses in the manufacturing sector will be offset by gains in the service sector. This belief is buttressed by the neoliberal doctrine that the market will decide the best mix of sectors; in this view there are 'no grounds for regarding any particular sector as more fundamental than any other' (Brittan, 1985).

These claims can be challenged on a number of counts. Although manufacturing now only accounts for around 25 per cent of total employment in the UK, it still accounts for some 75 per cent of the UK's visible export trade. For services to compensate for this the UK would have to increase its share of world service exports from less than 10 per cent to over 50 per cent, an impossible task. In fact, the contrary seems to be the case because, in aggregate terms, the UK service sector has fared no better than its manufacturing counterpart as regards its share of world trade (Bank of England, 1985). What is more, the fortunes of these two sectors are becoming more interlinked: manufacturing is not only an important market for producer services, but the value of services 'farmed' on the back of manufactured goods is increasing rapidly. As for the claim that the market can be relied upon to decide the best mix, this is both naive and dangerous. It is naive because it was government, not the market, that created the absurdly high exchange rate in 1979–80 which gutted both jobs and capacity in manufacturing, and it is dangerous because it ignores the extent to which high-tech markets are increasingly being structured by political forces.

It is not difficult to see why so much of manufacturing industry feels it has been 'written off', or why there has been a strong demand for a state-led industrial strategy (House of Lords, 1983; Engineering Employers' Federation, 1987). What is unique about Britain is that it is alone among the major OECD countries in suffering an absolute decline in manufacturing output since 1974. So much so that after seven years of the neoliberal experiment manufacturing output was still lower than it had been in 1979, and a manufacturing trade surplus of over £3b in 1980 had deteriorated into a deficit of over £8b in 1986. Britain, it seems, is becoming less and less a 'workshop of the world' and more and more a consumption centre for luxury consumer goods and sophisticated engineering equipment which are imported from abroad. Quite apart from the obvious balance of payments problem, itself obscured by huge but declining oil revenues, what we see here is a growing material disintegration of the national economy. Because a country's innovative capability is heavily dependent upon a dynamic pattern of stimulus and response running vertically through customer–supplier linkages within the national economy, it seems that Britain's deteriorating trade position is also weakening its capacity for industrial innovation (Radice, 1984; Organization for Economic Cooperation and Development, 1985).

Although industrial profits have steadily increased since the trough of 1981, there is little visible sign that they have been used to address the weaknesses identified earlier. For example, only a tiny fraction of the £3b relief on national insurance surcharges appears to have been ploughed back into company R&D; not surprisingly, total R&D employment in Britain has been steadily declining since 1980. The Thatcher Government is not beyond admitting its fears on this score; in fact it has betrayed 'a quite profound disenchantment with industry's response to government policy', especially with respect to the lack of investment in R&D, training and the like (*Financial Times*, 1986), but then the neoliberal strategy has done little to enable it to respond.

This reading of the situation may seem unduly pessimistic in view of the cyclical recovery of recent years – a recovery fuelled by what hs been called the 'fall of monetarism', i.e. the abandonment of strict monetary control, the large devaluation of sterling in 1986 and a pre-election spending spree (Smith, 1987). But, for all the official talk of an 'economic miracle', the fact remains that UK growth between 1979 and 1986 was average by Western European standards, below that of West Germany and well below that of Japan and the USA. Indeed, if oil production is excluded, the UK growth rate for this period falls to near the bottom of the league of major OECD countries. Furthermore, a series of critical indicators – civilian R&D, current skill levels, training provision, education, capital stock and international trade – signify not a strong but a dangerously weak foundation for a sustained recovery, let alone a genuine renaissance (Godley, 1987).

Yet, for all this, Thatcherism has been remarkably successful in sustaining itself politically, and there are at least three major reasons for this. First, the heaviest costs of the neoliberal strategy have been borne by areas outside the Tory heartland of southern England. (For example, the three southern English regions accounted for a mere 6 per cent of the UK's total job losses between 1979 and 1986. These southern English regions seem to be becoming increasingly uncoupled from the rest of Britain, both economically and politically. No modern Conservative government has ever been so electorally dependent on southern English seats – 61 per cent of all Tory seats were drawn from this southern bastion in the 1987 General Election. The second factor sustaining Thatcherism is that those in full-time employment have been cushioned, albeit unequally, by rising real incomes. Against this, however, Britain's 'underclass' – defined as those living on or near to supplementary benefit income – has increased from 6.1

million in 1979 to 11.9 million in 1986 (see Rentoul, 1987). Indeed the very notion of a 'recovery' is an affront to such people. Finally, there is the simple but fundamental fact that the political opposition to Thatcherism is hopelessly divided. So, unless recent electoral trends are reversed, the policies required to engineer a socially progressive and spatially balanced recovery will continue to be politically vetoed by the more affluent social strata of southern Britain.

PART IV

Theoretical and Policy Conclusions

14

Conclusions
In and Against Uneven Development

We shall conclude by bringing together some of the main theoretical issues arising from the work and discussing some of the implications for industrial strategy or policy.

In contrast with the popular view of the electronics industry as a cure for economic problems, we have argued that it is no less subject to uneven development than other capitalist industries; indeed, given the rapidity of technological change in electronics, its development tends to be more turbulent than most. This is true not only of semi-conductors, with its spectacular cycles, but increasingly of formerly settled sectors such as telecommunications. This dynamism, coupled with the highly internationalized character of much of the industry, makes capital in electronics more mobile than in many other indus-tries. However, as we have shown repeatedly, it is still the rich coun-tries which dominate these flows, both as origins and destinations.

If the continuing role of markets as foci of foreign direct investment is one respect in which the *New International Division of Labour* thesis is deficient, another is in its underestimation of the extent to which the international division of labour is structured not only by corporate spatial divisions of labour of multinationals but by competition between, and different patterns of specializa-tion among, a large number of separate firms, many of which have significant degrees of spatial monopoly in certain countries.

Parallel to this is a common underestimation of the continuing importance of industrial agglomeration. The electronics industry is often cited as a prime example of decentralized production systems with branch plants having little economic impact on the areas in which they are located because their suppliers and markets lie elsewhere, i.e. the 'cathedrals in the desert' syndrome. While there are undoubtedly many cases of this, agglomeration is still important for the many types of non-routine production requiring close buyer–supplier interactions; indeed the need for these seems to be increasing (see IBM, 1984; Scott, 1985, 1986b; Glasmeier, 1986). Now in the traditional regional development literature (e.g. growth centre theory) the latter kind of situation was, of course, given excessive weight; studies of multinationals' branch plants showed that they made much less use of local sourcing than did small indigenous firms. But as with the concept of the new international division of labour there is now a danger in overreacting against the traditional assumptions, again because of the obsession with mass production at the expense of other activities more likely to use local sourcing.[1] Yet, even in mass production, there are now signs of a re-evaluation of the costs of decentralized production in the light of the success of Japanese firms in achieving continuous innovation through permanent interaction between development work and production. By contrast, treating production as mature or standardized and moving it out to a low-cost location has the effect of choking innovation by separating production from development.

One of the consequences of combined and uneven development is that spatial mismatches between causes and their effects become the norm. The fate of workers in a particular locality comes to depend on the nature of much wider structures within which they operate. We saw how the nature of the industry within Britain and its regions depends strongly on its place within the international division of labour and in relation to markets of different degrees of wealth and technical sophistication. The fate of workers in South Wales is linked to that of certain workers not only in the M4 corridor but in places as diverse as Silicon Valley, South Korea, Tokyo, the Mexico–US border and many others. And they are linked not only through the market but often as parts of the same international corporations. Now they are also organized within the same firms, making local economies increasingly sensitive to the play of market forces on a global scale. Conversely, international production systems depend on what is happening in the particular localities in which they are rooted. Even in the case of foreign direct investment, success depends not just on what is happening in the host region or country but on the character of the home country. Though international capital is increasingly mobile, it is far from rootless; and putting down new roots is far from straightforward, especially where the goal is to penetrate foreign markets. The development of an industry such as electronics therefore has to be understood in terms of the articulation of processes operating and manifested on a variety of different scales. The dangers of ignoring this combined and uneven development are most evident in policies at local and national levels for regenerating industry, whether they involve attempting to clone areas or countries with unique situations, such as Silicon Valley, Berkshire or Singapore, or cloning state policies or institutions tailored to unique contexts, such as MITI.

Despite the global character of the industry, nation states and their governments are significant actors in the process of uneven spatial development from the start. The successes of Silicon Valley etc. – places which neoliberal rhetoric and the ideology of high tech celebrate as the products of free market forces – are in fact heavily underscored by major state investments. However, it is not enough merely to say that here, as always, 'the capitalist state supports capital accumulation and the interests of capital'. The attempt of each government to support 'its own capital' stumbles across an increasingly tangled web of international competitive and collaborative relationships, as we saw in chapter 7. And, given the complex game-like structure of competition and the related unpredictability of market and technological trajectories, it is not surprising that what policies actually favour 'capitalist interests' is itself ambiguous. 'Support' can weaken, as well as strengthen, a national industry, as a comparison of British and Japanese kinds of 'state support' shows.

While many of the above theoretical points may seem relatively uncontentious, some readers might be concerned about our refusal of reductionism and our emphasis on detail and differentiation. Normally, economic analyses focus on the general operation of the calculus of exchange value and some may wish to dismiss our concern with the particular material forms that capital accumulation takes as quibbles about 'different ways to skin a cat'. But abstractions need to be combined and supplemented to make sense of the concrete world and reductionism is a good way of misunderstanding capitalist industry, the way it is changing and the experience of workers. Moreover, as we shall argue shortly, these complexities massively increase the problems of displacing capital; ignoring them will not make those problems go away. Nor does our approach amount to a retreat into a thesis of radical uniqueness. Such uniqueness or individuality as exists is formed through interdependencies with other places, policies or institutions, and through the operation of general processes such as product innovation and capital accumulation. This kind of paradox is typical of combined and uneven development; in some respects progressive internationalization homogenizes global space, reducing it to the same standard of economic calculation and reducing isolation. Yet the 'annihilation of space by time' also allows places to become *more* individualized and differentiated, since improved communication has made duplication less necessary. On the other hand, some phenomena can be broadly imitated or at least adapted in different contexts, even where the original circumstances were unique. This is particularly clear with management-labour relations, where, in a quasi-evolutionary way, conventional practices are adapted to local circumstances, sometimes leading to the development of higher forms which turn out not only to be more vigorous than their forebears but equally imitable. But it is also evident in something we normally take for granted – the fact that capital can continually take root in new environments. In this sense, local developments are continually diffused globally.

However, we believe the theoretical implications of the fore-going chapters go deeper than a critique of reductionism and the underestimation of uneven development. Although strongly influenced by Marxist theories of capitalist industry, our empirical analyses and our engagement with business literature have led us to query dominant interpretations of those theories at a more basic level. The problems centre on the interlocking relationships between capital (C) and labour (L), and between capitals, as represented thus:

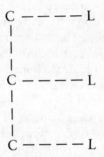

On the face of it, there is nothing controversial in identifying such a structure but we wish to argue that, by paying insufficient attention to the capital–capital side, many Marxists have inadequately understood the structure as a whole. The symptoms of this problem are many: the neglect of the situation of firms within the social division of labour, the neglect of product innovation as a competitive strategy, reductionist views of markets, and certain misinterpretations of management–labour relations and of spatial divisions of labour. Complementing these areas of neglect, and reflecting the political preoccupations of the Left, there is a tendency to read most eventualities as outcomes of the changing relationship and struggle between capital and labour.

Insofar as orthodox Marxist accounts deal with inter-capitalist *competition*, they do so overwhelmingly in terms of the capital–labour relation. So, competition is seen both as the whip hand forcing individual firms to extract more labour from their workers and as a consequence of firms succeeding in doing just that. We readily agree that this is an important, perhaps the most important, process in the interaction between the capital–capital and the capital–labour sides of the structure, and indeed we have documented examples of new forms of this process, particularly in volume production industries such as consumer electronics. However, the two sides of the structure also have a significant degree of autonomy from one another. Product innovation is often obligatory for survival and the need for this is largely independent of the state of firms' capital–labour relations, though of course it may subsequently have a large impact on labour. It is also true, as chapter 12 in particular showed, that developments in capital–labour relations, especially regarding professional and managerial workers, can be directed principally towards facilitating continual incremental product innovation or modification, though this is rather different from trying to push up labour productivity in the production of a fixed range of products – the case that the Left traditionally focuses upon. We further readily concede that some product innovations are a response to the need of user firms to introduce process innovations, perhaps in some cases for controlling labour better. Recent Marxist accounts tend to use this as an excuse for reducing the situation of the supplier firm to that of the user firm, so that again technological change is seen purely in terms of the capital–labour relation. Our point is that firms need to be seen as technology producers as well as technology users: the characteristics deriving from the former cannot be read from those deriving from the latter. This reminds us of our point, implicit throughout parts II and III, that firms face *other* problems besides those concerning their relationship with labour, in particular those involved in responding to changing product technologies and markets, and these in turn are not simply reducible to consequences of successes or failures in disciplining labour. Thus, while any kind of restructuring affects the capital–labour relation, it is often done for other reasons. The restructuring carried out in firms producing telephone exchanges with the shift from electrical to electronic equipment is a good example: although it caused hundreds of thousands of job losses in Europe and North America it was not done for this reason. And although it created the opportunity for changing to new, 'green' workforces, the restructuring would have been necessary even without this opportunity (see Thompson and Bannon, 1985).

Stories of instances of restructuring undertaken for reasons of process efficiency and control over labour have now entered the conventional wisdom of the Left. The political implications of this approach are generally seen in terms of the need to strengthen labour against capital and its management, so that capital cannot pit one group of workers against another to their mutual detriment. Yet, where this does occur, 'capital' divides and rules because it is itself divided and in continual internecine conflict. Our response is therefore not to deny that such restructuring occurs, or to reject the political goal, but to stress that it is far from the only process by which capital accumulation proceeds. However, it is because of the one-sidedness of the orthodox view of restructuring that its proponents can say so little about what a strengthened labour might *do* with industry.

Complementing these biases is a striking neglect of markets and the problems faced by capitals in creating, penetrating and coping with them. In what might be termed the 'productionist tendency' it almost seems as if commodities sell themselves or as if all markets were like those for apples. The neglect may support or be supported by the illusions that in a socialist society the market could be completely supplanted by 'The Plan', but this is nowhere more unlikely than in an industry as complex and changeable as electronics. More simply, the problem reflects an astonishing underestimation of the complexity of capitalist economies; as we noted earlier, while we cannot expect abstract theory to anticipate the form of the social division of labour we should at least expect it to alert us to the fact that it is enormously complex and that this in itself has major economic and political consequences. In other words, the orthodox approach reduces the problems that capitals face in maintaining a place in the social division of labour to the problem of minimizing the labour time, or other production costs, of producing a familiar, standardized commodity whose marketing is unproblematic.

The implication of these points for particular groups of workers is that their insecurity derives not simply from being propertyless. The division of labour within most manufacturing firms, between a minority responsible for gaining a place in the social division of labour (marketing and part of management) and a majority restricted to the details of production, reinforces the difficulties the latter have in determining their destiny. Yet, even in the absence of such a division, their insecurity derives from belonging to an organization competing within and dependent upon a wider social division of labour. The latter is governed in turn by the law of value and mediated through the largely uncontrolled and uncontrollable drift of technological and market trajectories. Job losses through restructuring may therefore often be attributable to a failure of employers to keep up with changing product markets and technology although, clearly, not all can keep up or keep ahead simultaneously.

Our criticisms of orthodox treatments of spatial divisions of labour at both the international and intra-national scales (recall especially chapter 7) echo these problems in three respects. Firstly, the spatial divisions of labour tend to be seen too little in terms of the social division of labour – roughly corresponding to the capital–capital side of our structure – and too much in terms of technical divisions of labour, which of course relate closely to capital–labour relations within firms. Secondly, corporate, technical spatial divisions of labour are seen primarily as reflections of capitals' relationships with different kinds of labour rather than sometimes as by-products of market strategies. So, for example, radical analyses of the newly industrializing countries tend to emphasize state repression of organized labour, super-exploitation of cheap labour and the position of branch plants within multinational technical divisions of labour and then attribute their development to this. Yet, while the conditions in what Lipietz terms 'peripheral Fordism' certainly need to be exposed, they do not amount to a sufficient condition for the rise of the newly industrializing countries, for many countries with cheap super-exploited labour show little sign of catching up with more developed countries (Lipietz, 1984). What matters is not only the state of capital–labour relations but the position of capital in those countries with respect to technological and market trajectories, the selection or targetting of particular products and sectors, backed up by long-term state support, and the relationship of indigenous capital to foreign capital in terms of subcontracting, licensing, joint ventures, technology transfer and so on; the Korean case illustrates this perfectly. 'Success', in other words, depends on strategy with respect to capital–capital relationships too.

While there are many cases where the orthodox radical kinds of explanation seem appropriate, their limitations can most easily be seen from a different, though possibly overlapping, perspective which emphasizes capital–capital relations in terms of competitive strategies, markets and innovations, especially in products. This alternative perspective is dominant in the business literature and has become increasingly influential in academic interpretations of industry. It is primarily concerned with the challenges facing companies in their markets and with their corporate strategies. Restructuring is seen primarily from the point of view of capital rather than labour and uneven spatial development is seen more in terms of inter-company competition, unevenness in the development of markets and partial spatial monopolies than in terms of spatial technical divisions of labour.

The role of labour tends to be underestimated in this approach and from a Marxist point of view it is therefore tempting to dismiss it. However, its strengths correspond to the weaknesses of the radical approaches (and vice versa) and it therefore provides some elements of a corrective to the problems described above, particularly with regard to management and markets and the 'productionist' tendency. While Marxism rightly points out that capitalist management is about social control and exploitation and not just about technical and organizational matters, in recent years it has tended to forget the latter side, thereby giving the impression that when workers took control management would not be a problem. Those problems of production not directly related to disciplining labour are passed over or reduced to ones of labour control (but see Tomlinson, 1982). So, for example, we noted in chapter 12 that the problem of the 'software bottleneck' tends to be seen by the Left as stemming from the control that programmers have over the labour process, not as a problem deriving from the inherent difficulty of producing software. Nevertheless, while many of capitals' problems do not derive from labour they certainly impact on labour. Therefore to dismiss such issues on the grounds that we are interested in labour's problems, not capital's, is myopic, not to say undialectical. And if one is concerned about alternative forms of economic organization to those of capital, then ignoring these problems of management is a recipe for failure. A further consequence of this neglect of non-labour-related problems is that it reinforces another common misperception on the Left: the view of large firms as monolithic and omnipotent. Ironically, this can lead to an underestimation of their domination by the law of value and of the uncertainties of devising product and marketing strategies which enable them to keep their place in the social division of labour.

Our interpretation of the interaction between firms and *localities* again depends on how the capital–capital/capital–labour structure is understood. We have seen many examples of how regional characteristics, especially the nature and prior socialization of labour, affect the competitiveness of the firms within their boundaries, not just in terms of the cost of inputs – particularly labour power – but in terms of how those inputs can be used. Yet, while there is this local influence, there is also a deep divide between the interests of the firm and those of the workers in the region or regions in which it is situated. The relationships or interdependencies between firm and region are highly asymmetric for there are other, often more critical, determinants of company performance which have nothing to do with the region. In other words, the region depends on the firm more than the firm depends on it. Consequently, each region is at the mercy of wider changes in product markets and international competition over which it has no control; a region and its workers can hardly be blamed if automation or a superior product from Japan puts its local electronics plant at risk. Now, at one level, these points are widely recognized, but if competitiveness is read entirely as a matter of process efficiency then there is a risk that we may actually *overestimate* the importance of local variations in workforces, simply because we underestimate the extent to which non-labour factors matter. In fact, the Left may inadvertently provide ammunition for the Right in blaming regions for their own regional problems.

From the point of view of labour the asymmetry between firm (or plant) and region can be seen in terms of minority (often external) control of the means of production which puts the fortune of a substantial proportion of the local economy beyond the control of local people, even where they are employees of those firms. In addition, local people may have little knowledge of, or interest in, the arcane details of their local computer company's operations and its changing technologies and markets, even though their livelihoods may be strongly affected by its performance. Part of the reason for this may be that few local people occupy an elevated position within the vertical and technical divisions of labour within the firm. But it is also an effect of the firm's highly specialized role within the social division of labour and the inevitable dependence of society at large upon particular bodies of specialist knowledge and skills which it is only feasible for a small minority to grasp. In other words, the asymmetry cannot be understood simply in terms of class; the related dimension of the social division of labour and capital–capital relations is crucial.

These claims can best be assessed by considering what would change in the event of worker takeovers of firms, of expropriation of the owners of capital. The most elementary point in socialist economics tells us that on its own this would change little, since the workers would still have to find a place in the social division of labour, large parts of which are too complicated to plan, and hence would have to survive in competition with other firms, whether indigenous or foreign, whether run by capitalists or workers. If this is such an elementary point in the theory of socialism, why do our theories of capitalism so often fail to anticipate the point by giving the impression that the insecurity of workers' livelihoods derives only from their class position?

We are well aware that the general thrust of our argument carries with it the implication that labour and class struggle have had less effect in shaping development in the industry than is generally believed on the Left. In most cases, labour has had little opportunity to take the initiative in forcing capital to change in a certain direction, not least because the rate of technological change has been such that the parameters of concrete capital–labour relations change so frequently. On top of this there are enormous problems in organizing internationally fragmented workforces.[2] This is not to say that resistance has been absent, but that on the whole it has been reactive rather than pro-active. While this may play down the role of resistance, it need not belittle the importance of agency, for intentional action is a necessary condition of everything we have reported, be it transformation or reproduction of the *status quo*.[3]

In part, the overestimation of the role of class struggle derives again from the one-sided treatment of the capital–capital/capital–labour structure, coupled with the usual dose of wishful thinking. But there are other roots to the problem as well. One is a neglect of the extent to which (objectively) all production involves cooperation as well as conflict and in turn the extent to which (subjectively) the former is often given willingly rather than reluctantly.[4] Management by consent and cooperation are therefore not to be explained simply by reference to ideological control, but as having a genuine material basis (see Przeworski, 1980).

This is related to a common failure to analyse the relationship between workers and their employers in concrete as well as abstract terms. At the most abstract level, we can identify the fundamental conditions and contradictions which are present wherever capital and labour exist, regardless of the particular concrete forms they take. Concrete analyses such as our discussions of company performance and management–labour relations can recognize that the practices are structured by the deeper relationships isolated by more abstract research. At an abstract level, the interests of workers *as such* are opposed to those of capital in terms of the production of surplus value and the fact that workers are not allowed to produce for their collective needs and in ways that they themselves control.[5] But this does not render some degree of 'humanization' of work impossible.

There are, then, differences in the interests of workers in general and particular groups of workers. Notwithstanding their class position, the latter inevitably have an interest in the success of their firms, at least as an alternative to unemployment, even though success of the firm may not guarantee their job security and despite the fact that it may weaken the position of competing capitals and hence threaten other workers' jobs. And, as we saw in chapter 10, employment growth can be a solvent of discontent and a facilitator of new management–labour relations in new branch plants.[6] On top of these divisions there are of course others too, particularly based on gender and occupation. All of which makes for an extremely complex picture: labour's interests differ across the industry and at different levels. Small wonder that the assessment of concrete labour and management strategies such as those described in chapters 10 and 12 is difficult.

In the concrete studies reported in part III we saw that two of the main contrasts in the electronics industry workforce are those between back-end and front-end workers and between workers in mass, standardized production and in small batch, customized production. As we saw when comparing the industry in South Wales with that in the M4 corridor, routine work is the more vulnerable to jobless growth, whereas increases in demand for front-end work such as development tend to require extra workers. But there are further implications of this kind of contrast in relation to the law of value and uneven development. When we speak of the law of value we mean more than simply competitive and market forces, for the theory of value on which the law is based centres on the link between competition and exchange on the one hand and labour and the labour time expended in producing commodities on the other. The relationship is most obvious where a leading capital undercuts its rivals by reducing the labour time involved in making its products, as the Japanese did in television production. Where, as in this case, the product is mass produced and in competition with very similar products, the effects of such developments are felt by workers in both the innovating firm and its competitors in a highly direct fashion, since price competition is keen. Where there are niche markets and product innovation and differentiation play a bigger role, the workers are less exposed to the law of value because buyers do not discriminate so keenly on the basis of price, and therefore differences in labour time among competitors are not so critical. Since particular custom products and services are not already on the market, buyers have limited information on which to base choices between competing suppliers. And, precisely because such commodities have yet to be produced, the quantities of labour time to be used in making then are not nearly as predictable as in routine work in mass production. In addition, non-routine work inevitably allows more idle time, as many professional, clerical and service workers know well. The nature of the work is therefore not only less routinized but generally less pressured. This may not be true, however, of development workers where they have to develop custom designs for commercial markets (e.g. computer systems) for here deadlines are frequently exacting and failure to meet them may result in lost sales in the future. Nevertheless, where deadlines are distant and competition weak, as in some parts of the defence industry, development workers may also enjoy a more relaxed work regime.

In the electronics industry the less-routinized forms of work tend to require a larger number of skilled workers and of course better paid ones than in more routinized work. But the contrast between the front-end (overrepresented in Berkshire) and routine back-end work (overrepresented in South Wales) and its effects on work culture, and presumably on political consciousness, goes further than their different occupational profiles and associated contrasts in incomes. What we are suggesting is that the different degrees of insulation from the law of value within the industry also result in differing experiences of work and that these too may have important effects on political consciousness. Further, if, as seems likely, these kinds of work are more common in other sectors, particularly outside manufacturing and in the public sector, this may prove a significant source of social and political divisions, perhaps with a distinct spatial expression too as the proportions of the different kinds of work vary.

Some strategic issues

The theoretical and empirical conclusions we have discussed all have implications for industrial strategies and for who might benefit from them. In this final section we wish to raise some of those implications, not to propose the details of a strategy ourselves but to clarify some of the issues which would have to be taken into account in formulating one, particularly those regarding the scope and limits of state action.

No industrial strategy is distributionally or politically neutral, though many pretend to be. The neoliberal scenario happily subordinates the interests of labour to capital although, as we have seen, it is far from ideal for many sections of capital. The ideology of high tech, with its blinkered pursuit of new technology for its own sake, marks an abandonment of economic and political thought. Against these stances, we insist that we have to consider what we want electronics or high technology *for*, and whose interests it favours or might favour. These questions in no way presuppose a Luddite position, nor are the answers simple and self-evident. What the Right takes to be natural laws of the market economy are in fact politically contestable social arrangements. Efficiency and democracy are generally accepted to be desirable features of social organization and yet there are those at both ends of the political spectrum who assume that they are incompatible in economic organization. We believe that the incompatibilities are less pervasive than is commonly thought. Hence our interest in informing thinking about *socialist* industrial strategy is based on the view that there are ways of increasing democratic participation which can enhance, rather than limit, efficiency.

With the exception of military electronics and telecommunications, the combination of a high degree of internationalization of production and markets with unparalleled technological dynamism has meant that the pressures of the law of value in electronics have been extremely powerful. Moreover, in Britain, we start from a position of competitive weakness, technological dependence and foreign penetration and thus have little influence over the trajectory of the industry. Consequently, no alternative strategy could hope to extricate the industry from this context and simply escape the pressures of the law of value and the uneven development it engenders. Uneven development is not a local aberration, removable by appropriate state policy, but is endemic in capitalism. Nor can simple 'boosterism' of the electronics industry be presented as the answer to problems of uneven development, (a) because it is only one industry, not the whole economy, and (b) because there is little reason to suppose that today's modern industry should be any less responsible for reproducing uneven development than earlier ones. Hence our title – 'In and against uneven development'. Point (a) serves as a reminder that this cannot be a discussion of socialist strategy in general; (b) requires amplification regarding the extent to which one could hope to counter the law of value and the forces of uneven development.

The sheer complexity and dynamism of the sector severely limit the scope for both centralized planning and bottom-up grass-roots control. Even worker-controlled firms, sponsored by the state, producing 'socially useful products', would be obliged to enter into a large number of market transactions and would have to pay off their creditors, including large capitalist firms in control of advanced products which they needed. In most cases they would find themselves in competition with other firms, whether near or far. For these reasons, unless special financing and contracts were arranged (e.g. though public procurement), they would not escape the pressures of the law of value and would have to behave in a basically capitalist manner. In general, then, all we can suggest are ways of 'bending the stick'.

What room for manoeuvre, then? There are three main kinds of possible influence:

1 over the survival and growth of firms (and employment) in the market;
2 over the development of markets and the definition of needs;
3 over the conditions of work inside the firm and the control of production.

Point 1 is essential, but while it may seem an obvious point it is often overlooked that survival and development require continual innovation in products, processes and management. If markets are not expanding and diversifying, the success and failure of firms which are competitors are always interdependent and relative; the success of firm A in this context entails the failure or backwardness of competitor B, and the workers of both are likely to suffer – hence the importance of encouraging the enlargement and diversification of markets, a goal achieved via both point 1 and point 2. However, point 1 on its own need have no socialist content. It does not even guarantee employment growth, though it at least makes it probable at the level of the economy as a whole, whereas an inefficient, technologically backward and uncompetitive industry bodes ill for labour, whether as producers or as consumers.[7] Chapters 10 and 12 should have established the importance of managerial innovation within firms but there is also scope for organizational innovations in inter-firm relations for, as the Japanese experience shows, industrial efficiency and performance can be boosted by this. For example, the innovative capacity of large Japanese firms, and their lack of worries about short-term fluctuations and possible takeovers, depend heavily upon forms of organization between firms and relationships with financial capital.

Point 2 acknowledges our opening point that we have to consider what we want new technology *for*; markets do not arise out of nowhere but are created, and they can therefore be politically influenced. The prime examples in electronics are the markets generated by military and other forms of public expenditure (of which more later), though other redistributive policies could be considered beyond the existing reach of the state. Here state macro-economic policies can of course affect both point 1 and point 2.

Point 3 is the most clearly related to traditional socialist concerns and involves the position of the industry's workers, not only their conditions of employment but their influence upon their own destinies. Attempted improvements in this sphere have little hope of success if divorced from points 1 and 2. The Left-controlled Greater London Council's activities and literature are interesting in this respect for, in concentrating on point 3, they extended beyond the traditional concern with working conditions and employment policy to the control of production itself, hence challenging power relations within the firm.

The most difficult questions concern the relationships between points 1 to 3: are they necessarily contradictory or possibly harmonious? It is often assumed on the Right that strengthening labour's position at work means making capital less competitive and similarly on the Left it is assumed that making capital more competitive means weakening labour. This may indeed happen, particularly in Britain, where technological and managerial backwardness invite capital to take the easy option of competing by cutting wages, using new technology for displacing labour rather than developing products and services and withdrawing from competitive markets; the traditional management views of skills and industrial relations that we noted in chapter 10 have much to do with this. Yet, there and in chapter 12, we saw that 'enskilling' rather than deskilling could work in capital's favour and that good working conditions and pay could facilitate success. This suggests that the relationship between outward performance and internal work conditions (in the broad sense of point 3) is more complex than is often supposed. By the same token, the insulation from market pressures enjoyed by many public sector workers does not necessarily mean that they have more satisfactory and satisfying working conditions. In other words, there is an area of contingency or indeterminacy in the relationship between the profitability and performance of firms in their markets on the one hand, and the conditions of work inside the firm on the other. Taylorism is one way in which firms have tried to improve their competitiveness, but it has its limits and is being superseded in some quarters. 'Post-Taylorism' usually involves an intensification of labour, but whether this makes the subjective experience of work worse or better depends on many other conditions, as we saw in part III. Our point here is that this contingency creates some scope for labour to exert greater leverage without worsening the efficiency and effectiveness of production.

One example of this contingency concerns the possibility of making collective production more egalitarian and democratically controlled without making it hopelessly inefficient. While we hold no illusions about the dispensability of hierarchies, a labour strategy could challenge both the conventional allocation of particular types of people to particular jobs and tasks (e.g. white males only in high-status jobs) and the conventional ways of dividing up the tasks and responsibilities themselves. None of this contradicts our earlier point that in certain fundamental respects the interests of labour and capital are at odds, for capitalist production involves cooperation as well as conflict and exploitation, and positive-sum elements as well as negative-sum elements.

All three areas of influence discussed above require detailed knowledge of the material conditions within and outside firms. Throughout this book we have tried to show how the material forms of industry – both technological and organizational – are crucial to its development. Against this view we have the unlikely bedfellows of neoliberals and ultra-leftists for, despite their obvious opposition, they share the misconception that the material structure of the economy does not matter. On the Right this comes through in the lack of concern over the decline of manufacturing (and over the issue of what constitutes 'real jobs'). In this view exchange is mistakenly seen as wealth creating, as evidenced in the obsession with making money by buying and selling companies. Where material forms do make a difference, unfettered market forces (i.e. capitalists responding to market signals) will sort things out in the best possible way. At least in some parts of the far Left, the crisis is seen as caused by a profit squeeze which itself does not relate to any particular organizational and technological forms (e.g. Armstrong et al., 1984). All that matters is value and the struggle between capital and labour over this. Of course, capital *does* have to prioritize exchange value, given that profit is the bottom line. But the success of companies in meeting this obligation depends on their performance on the use-value, material side.

Views on these matters in management circles have actually changed in the recession. In the long boom, 'IT&Tism' had become popular, in which firms were no longer restricted to particular sectors by virtue of their resources and accumulated expertise but became umbrellas for activities as diverse as telecommunications and hotel management. The popularity of growth by acquiring whatever businesses were profitable was symptomatic of the low priority given to the use-value forms of capital and to hard-won in-house technical and market knowledge. However, under the rigours of recession, such strategies have foundered, and now there are signs of a renewed respect for such expertise and technical capacities: money capital may be highly mobile but the material and intellectual conditions of production are not. This is evident in the recent 'back to basics' movement, in which companies divest themselves of all but a core of activities in which they have established expertise.[8] Similarly, the problems of convergence in information technology (IT) illustrate the importance of the material and organizational bases of accumulation as does the increasing popularity of inter-firm collaboration as an alternative to simple takeovers. And the success of Japanese firms has much to do with the way in which short-term exchange values are not allowed to obstruct the development of advanced production capabilities over the long term.

A further misconception shared by both neoliberal and ultra-leftist schools is the treatment of the economy as an ocean of market forces, studded by islands of organization in the form of capitalist firms. Non-market inter-firm relations and issues of industrial organization are consequently ignored. Yet it is clear that uneven development is significantly influenced by differences in forms of industrial organization – most obviously in the contrast between highly organized industrial groupings in Japan and the weaker, more purely market-based, relations found in Western firms. Industrial organization matters and is increasingly seen as crucial for competitive performance and any effective industrial strategy would need to be aware of this. Yet what is important regarding industrial organization is not so much formal questions of vertical integration or disintegration defined simply in terms of ownership or, indeed, in terms of firm size. For example, while new, initially small, specialist firms have figured prominently in major new technologies in the USA, it is certainly not impossible for large vertically integrated companies also to be highly innovative, as Japanese electronics firms have demonstrated. What matters in both intra- and inter-firm relations is whether actual organizational practices encourage innovation and efficiency and they can do this within a variety of different structures of ownership.

Paralleling this emphasis on organization is a similar stress on management, not just in terms of its inward-looking side, organizing production within the firm, but also in terms of its outward-looking side, concerning its marketing and its relationships with suppliers. Management cannot be reduced purely to a feature of the class nature of capital and hence labour cannot hope to gain positive, as opposed to negative, power without taking the issue of management as a whole seriously. Nor indeed can it afford to neglect the roles of professional workers doing work such as product development and marketing. Power in production *and* marketing depends not only on the stereotypical manual working class but on all parts of the technical division of labour.[9] Generally, then, traditional Left preoccupations have to be modified. In principle there is no contradiction between (a) wanting to strengthen the position of the weakest in the labour force, (b) recognizing that the tasks carried out by the strong are important too, and (c) challenging the division of labour between the two groups.

As we saw in chapter 13, state support of the electronics industry in Britain has been both weak and misdirected. However, the support programme outlined by the National Economic Development Office (NEDO), with its emphasis on strategic planning and long-term funding allied to enhanced support for civilian R&D, education and training, does go some way towards goal 1 – securing a viable future for the indigenous electronics industry. Sadly it has been spurned by the Thatcher administration. Given the short-term proclivities of the market in the UK it is difficult to see how its key supply-side weaknesses can be redressed other than by state-sponsored initiatives.

On the demand side public procurement is the single most important conduit through which the state can exert its leverage to induce and diffuse innovative products, especially in the field of IT where the public sector is the largest single user. But for a number of reasons this power has never been fully exploited. Firstly, it is so fragmented across a multitude of public agencies that its effects are dissipated. Secondly, low public investment in IT has prevented the public sector from playing a vanguard role: in 1986, for example, there were only four computer terminals for every 100 employees in the public sector, compared with 24 per 100 in the banking sector.[10] Thirdly, public agencies have paid too little attention to world export markets when specifying equipment, while too much emphasis has been placed on low initial cost as opposed to whole life cost (Advisory Council for Applied Research and Development, 1980). If these trends can be reversed, the public sector could play an enormously important role in fostering and sustaining an indigenous electronics industry in the UK. Without such concerted public sector support it will be difficult, if not impossible, to sustain a viable indigenous industry in a context where the purchasing power of the Pentagon alone is comparable with the entire purchasing power of the UK (Maddock, 1983).

However, if the potential of a state-sponsored strategy is not to burst on a bubble of false expectations, three major limiting factors must be acknowledged: (a) the fact that the national arena is no longer an adequate scale of reference for markets, R&D and standards setting; (b) that public procurement policies will lose much of their force if the recipients are allowed to wallow in a condition of state dependence; and (c) that even the most sophisticated innovation policies will come to grief if the indigenous company sector is not geared up to respond.

The first limitation might be construed as a blessing in disguise if the UK were to commit itself more forcefully to European initiatives. As it is, the UK is often perceived by its European partners as the Trojan horse of the EEC, too eager to sell itself as a base from which US and Japanese capital can penetrate the EEC market. Although there are severe obstacles to a concerted EEC alliance, as we indicated earlier, this option still seems to be the most hopeful route for the UK to follow. The fact is that without more effective European collaboration on R&D, standards, regulations and procurement, the current moves to create a more open EEC market will rebound to the benefit of US and Japanese firms because, unlike their EEC rivals, they already operate on a pan-European basis. Which products and firms to support should depend on long-term strategic needs and synergies, taking into account the EEC countries' starting positions within the international division of labour and markets and possibilities for collaboration at supra-national levels. At the same time, given the twists and turns of technological and market development, it is not advisable to put too many eggs in too few baskets, be they products or firms. Yet no country of even medium size can afford to support a full range of electronics products and, although the strategic significance of the industry is reminiscent of steel in an earlier age, the crucial difference for policy is that, unlike steel, electronics has been internationalized virtually from the start. Yet how far continued purchase of foreign technology perpetuates dependence depends on what is done with it – whether it is simply used, or whether it is used and adapted in a way which both improves the product and builds up the user's technological capacity.

The second limitation to state-sponsored strategies – 'feather-bedding' lagging indigenous firms – becomes particularly acute under 'patriotic' industrial policy regimes such as those operated by past labourist governments. In the absence of robust penalty clauses or competing suppliers, the UK telecommunications 'club' used its guaranteed public market not as a platform from which to export but as a substitute for exports (Morgan, 1987). Further-more, since the bulk of the output of the indigenous electronics sector is controlled by half a dozen large firms, upon whom a flotilla of small firms depend, the operations of these large firms ought to come under much greater scrutiny.

Often the benefits of state expenditure have not diffused beyond particular divisions of the large firms to their other divisions, let alone to their suppliers. Yet, as the experience of other countries shows, none of these problems are inevitable concomitants of state procurement and investment but can be counteracted by suitable sanctions and incentives. For instance, all the instruments at the state's disposal could be brought to bear on the task of discriminating in favour of innovative performance, so as to reward those firms engaged in heavy outlays on R&D and training etc. and to penalize those which are 'free riders'.

The third limitation revolves around the simple but fundamental point that no amount of state support can compensate for poor performance in the indigenous company sector, although NEDO's proposed package of support offers far more hope on this front than does that which exists at present. However, NEDO studiously avoids the fundamental issue here, namely how far the autonomy of the firm is taken to be sacrosanct if its performance remains poor. It is fundamental because it goes straight to the heart of capitalist relations of ownership, one corollary of which is that the shareholder is the only legitimate voice in the affairs of the company.

This raises one of the most contentious issues of industrial policy – the question of private or public ownership. Yet, despite all the rhetoric, with the exception of the issue of the appropriation of profit, many of the characteristics commonly assumed to be associated with either form of ownership are not in fact necessary features of them. Monopoly and its ills can occur in the private sector, as the newly privatized British Telecom (BT) shows. The profit criterion can guarantee that the needs of some – the poor – are *not* met. 'Short-termism' and insufficient size can cause suboptimal investment strategies. Public ownership need not rule out some competition. Whether it affords investment advantages is dependent on how the industry is run. Furthermore, public ownership does not necessarily yield control, let alone ensure that control will be exercised wisely.[11] While it is possible under public ownership to evaluate costs and investments on the basis of a social audit, this is not necessarily done. Whether it allows producer interests to overrule consumer interests again depends on how it is run, though in any case, as we argued earlier, the insecurity of workers' jobs is not purely based on their non-ownership of the means of production, nor are their interests as producers necessarily met by public ownership. But, while most of the ills alluded to above can occur under both forms of ownership, the scope for counteracting them is much greater under public control. So, while it can indeed afford certain advantages beyond the prevention of the private appropriation of profit, public ownership on its own is not a surrogate for industrial strategy any more than private ownership is under a neoliberal regime.

The economic case for selective public ownership in strategic spheres such as BT is therefore to be justified as a means towards an end rather than as an end in itself. With an annual procurement budget of some £2b and R&D facilities that are extensive by UK standards, BT is the single most important (civilian) actor in the UK IT sector. Although BT's procurement policy is still largely geared towards UK sourcing, it is less able to assume a 'flagship' role vis-à-vis the indigenous industry since privatization. For example, as a private company, subject to short-term commercial pressures, BT has little or no incentive to pursue strategic infrastructural investments such as broadband cable networks, capable of offering an array of interactive IT services to both residential and business users. Yet only BT has the scale and the expertise to construct a national network, as opposed to a patchwork of local networks confined to the more affluent areas. Such a national network would yield benefits not just for users but for suppliers as well. Another problem with BT's new private status is that its R&D facilities have become a proprietary resource, geared more towards short-term market pressures, rather than a national resource capable of being utilized for wider national and international collaborative ventures. Private capital has the power to subvert the wider potential of IT, as when *private* communication networks are used to lock customers in. This means that *public* access will become a more important issue – in this sense communication 'highways' are not at all like roads and railways, despite the ubiquitous 'transportation' analogy. Private ownership and control of the means of communication may therefore be a more pressing problem than private ownership and control of the means of production.

More generally, state ownership can allow a broader range of economic considerations to be taken into account in investment decisions than is the case under private control, where only costs and revenues borne by the producer have any influence. Most obviously, consumption by specific kinds of disadvantaged groups could be subsidized. However, the possibility of breaking free of capitalist economic criteria should not license a proliferation of subsidies with no attempt to count opportunity costs. As the Left sometimes forgets, some kind of economic accounting is still needed. Similarly, steps need to be taken to prevent producer interests overriding consumer interests, reproducing BT's reputation for appalling waiting lists and poor services to residential and business users and thus depriving the wider economy of innovative equipment, services and potential new employment.

The weakest part of the NEDO support package lies in the suggestion that indigenous firms should concentrate on their current strengths. Taken to its extreme this would mean a further retreat into defence markets, the major strength of most of the leading UK firms. The real issues are avoided if the debate is simply reduced to the question of doing more to increase the civilian spin-off from defence contracts, the situation at present. The chief problem with this approach is that, even if it were successful, it could only have a marginal effect on the fortunes of the civil electronics industry (Maddock, 1983). What is required, instead, is that the civilian sector receives the political commitment hitherto reserved for its well-endowed military counterpart, and that support goes beyond R&D. The importance of the latter point stems from the sequential growth of resources; for example, it is estimated that every £1 spent on research requires £10 at the development stage and £100 in production and marketing for a viable 'global product' (Wilmot, 1985). Short of a state-sponsored push in this direction, it will be difficult to undo the debilitating 'culture gap' between these two divergent sectors of the indigenous industry.

A strategy for the indigenous electronics industry clearly cannot ignore the growing presence of foreign multinationals. The knee-jerk response to this question by some sections of the Left is simple enough: they should be nationalized. But apart from the questionable wisdom of 'taking over' a foreign subsidiary which is but a part of a wider international chain, such a policy would probably encounter severe resistance from the workforce, since IBM and Sony, for example, have a better reputation for job security than ICL and Thorn-EMI. There is also the question of retaliation: a profound mis-understanding of how markets operate has blinded many on the Left to the fact that overseas sales in sophisticated product lines can only be accomplished on the basis of an overseas presence, as we argued in part II. This means that it is wrong to treat domestic and overseas expansion as if they are necessarily direct substitutes for one another.

So, if this is not a feasible option, what else can be done? Given the strategic nature of this industry, it is essential that inward investment policy is radically recast. The current uncritical attempt to woo international investors almost at any cost is reminiscent of a 'banana republic' approach.[12] Whether it involves front-end activities or branch manufacturing plants, inward investment is unlikely to have an import substitution effect; the effect is the opposite, as the UK IT trade deficit shows. It is not a question of shutting up shop to all forms of inward investment but, rather, of ascertaining how far this is consistent with policies designed to revive the indigenous sector. Otherwise, indigenous firms will continue to be displaced from current and potential product markets. Within such a framework there is a need for more rigorous monitoring of existing foreign establishments. The minimum aims here should be to ensure (a) that the effect on the balance of payments is at least neutral, (b) that local content and import substitution norms are raised, (c) that a spectrum of activities is created, not just routine assembly, (d) that job displacement effects are assessed and (e) that the right to bargain through a trade union (of the employees' choice) is fully recognized.[13] In some cases foreign firms already meet or approach these desiderata, but if these conditions are to be secured generally it will be necessary to win agreement at the EEC level to prevent multinationals playing one country off against another. Urgent cooperative action is required at this level anyway, because the competitive struggle for mobile capital has grossly inflated the level of national incentives and has thus led to an erosion of the tax base in each member state.

Even if the balance of power between international capital and the nation state is far from symmetrical, the nation state is by no means powerless, especially when several are aggregated together. If a concerted supra-national EEC strategy could be forged on conditions of entry then genuine progress is not at all fanciful. It is sometimes suggested that a critical stance towards internationally mobile capital is doomed because, without large national incentives in Europe, multinational companies will divert their activities to the lower labour cost locations of the Far East. But this underestimates the importance that US and Japanese multinationals attach to accessing the EEC market. Besides, as we have seen, the 'runaway industry' phenomenon largely consists of low value-added activities, which is not what the UK and its EEC partners should be engaged in anyway. To the extent that inward investment subsidies were significantly lowered or even abolished, the resources could be redirected towards funding new pan-European ventures and for strengthening support programmes beyond the collaborative R&D stage in existing European firms. The main point of these remarks is to counter the fashionable defeatism which asserts that little or nothing can be done against the grain of multinational power in this key industry.

Military electronics warrants special comment. Arguments about the proper role of military production can only rest upon arguments about defence policy. Yet just as extravagant defence projects like the USA's 'Star Wars' programme boost high tech, so the ideology of high tech invariably bypasses scrutiny of the ultimate purpose of the technology, thereby boosting the power of the military–industrial complex. Whatever one's views on defence policy, we must note the effect on civilian electronics and consider three further arguments regarding defence which are frequently overlooked:

1 Investing in defence is not like investing in consumer industries or intermediate industries. The use of chips in making televisions can make them better and cheaper. The use of chips in automating the production of means of production can make it more efficient and indirectly benefit the economy. Defence investments may be justified on other grounds, but in economic terms they are mainly a drain, particularly in Britain where civilian spin-offs are rare.

2 Military spending generates small amounts of employment at considerable cost. In its London Industrial Strategy the Greater London Council calculated from the 1984 Defence Estimates that the average cost per job related to strategic nuclear defence was £65,000, while for defence expenditure as a whole it was £31,000. By comparison the average cost of a job created by the Scottish

Development Agency was £10,000 per job in foreign-owned multinationals and £3,000 in small business, while the average cost of a job created by the Greater London Enterprise Board was £4,200 (Greater London Council, 1985, p. 297, para. 12.40).

3 The defence bias means that some of the most sought-after skills in the country and some of the best minds are devoted to the production of technologies of destruction. The obscuring of this predicament of so many advanced capitalist countries is perhaps the greatest achievement of the ideology of high tech.

Despite the major role of military electronics in the rise of the industry, the economic effects of cuts in military spending need not be detrimental:

> Both historical and econometric evidence suggests that disarmament raises few economic problems as long as three conditions are met. There must be a clear political programme for conversion, a compensating expansion in civilian demand to offset the reduction in defence expenditure, and appropriate supply-side policies to co-ordinate the transfer of resources to civilian use. In these circumstances, reduced military expenditure tends to increase employment, because defence spending is less effective at creating jobs than alternative forms of spending. (R. P. Smith, 1985)

State policy could also be made more sensitive to the *spatial* dimensions of the industry. It could restrain local and regional authorities from pursuing 'beggar-thy-neighbour' strategies and attempt instead to harmonize local, regional and national initiatives, while in procurement it could take into account the location of its suppliers so as to offset the bias towards affluent regions. However, it is important that such considerations come into play in decisions regarding *any* part of the country, not just those designated as assisted areas, and that spatial implications be considered in *any* state policy, not just those with an explicit spatial title, such as regional policy. Even then, mere geographical reorganization is not enough if all it involves is moving around inequalities rather than attacking them (see Massey, 1984).

Skills are of considerable importance both quantitatively and qualitatively: quantitatively, the shortage of skills during a time of unprecedented unemployment levels speaks of a massive shortage of training. Output of electronic engineers in the UK fell from 2,700 in 1984 to 2,550 in 1985 (interview). The USA produces 1.7 times as many electronic engineers per head as the UK, and Japan four times as many. No one in the industry doubts that serious skill shortages inhibit development but little has been suggested regarding the training shortage which causes it. While state-run training programmes can be useful, in-house training within firms is particularly important given the value of learning-by-doing, on the job. Further, in this kind of industry, user skills are at least as important as producer skills. A sophisticated supply sector cannot flourish without sophisticated users.

Industrial strategies traditionally connote policies which are both strongly centralized and state controlled; not surprisingly they are widely seen as dangerously bureaucratic and inflexible. These suspicions obviously have some foundation but, while it is important to seek ways in which the state can be made more flexible and responsive, there are several possible and desirable counter-tendencies to these characteristics. Clearly, an industrial strategy which is limited to a series of diktats and initiatives issued from the central state can expect to fall on stony ground. The problems an industrial strategy faces extend beyond the state itself and beyond macro and meso levels of the economy, down to shopfloor or office-floor level. One of the main implications of our research and of recent work on innovation has been the importance of cultivating learning-by-doing and expertise at the points where problems occur, often at this micro, grass-roots level, and the limitations of 'distance learning' in remote upper echelons of corporate hierarchies (Aoki, 1987). Such developments depend on more training at all levels, management included, and on reducing bureaucratic and direct control and giving workers more discretion; and although they do not guarantee an improvement in working conditions (in the broad sense) they do allow that possibility in a way which more Taylorist regimes do not. To a certain extent, therefore, altering technical and hierarchical divisions of labour in favour of workers need not be inconsistent with increased efficiency and effectiveness; in fact it could be favourable to them. Bravermanian critical theorists of Taylorism may find this a controversial point; if so, then perhaps they have become too much like their enemy. We suspect that few workers, whether blue collar or white collar, would argue with it.

Another related way of counteracting the bureaucratic and centralizing tendencies of state industrial strategy concerns innovation; it involves the state capitalizing on the advantages of a 'diffusion-oriented' innovation strategy over a 'mission-oriented' strategy, by decentralizing its support for R&D and training. By this we mean policies along Japanese, Scandinavian and West German lines of encouraging innovation in a broad range of users, including the public sector and so-called sunset industries, and of making government laboratories and training centres functionally and spatially close to them instead of remote and centralized (Freeman, 1987). Here, spatial decentralization is required less for its own sake – getting a more even spread of R&D workers and establishments in the hope of obtaining regional development benefits – and more as the best way of achieving effective innovation and development in general. Again, expertise is best developed near to the source of the problem.

While we consider that the above points are important for devising an industrial strategy, we recognize their preliminary and incomplete nature. Not least, there is the problem of political mobilization for, without strong support from labour – *including* professionals and progressive management – most of what we have raised is unlikely ever to materialize. But then political mobilization should in part be internal to the strategy itself, rather than an external factor, and we would argue that certain elements of our discussion meet that requirement.

This book has been about an industry which is the subject of more 'hype' than any other. While we recognize the industry's strategic significance, we end on a more sobering note. Most of the goods and services produced by the so-called sunset or low-tech industries still fulfil contemporary needs and provide much-needed employment in the process; industries do not age irreversibly but can be rejuvenated (or 'dematured' as the Americans say), not least by the adoption of new technology and work organization. Moreover, there are still unused resources in those industries and unmet needs for their products. The parameters determining the uneven fortunes of different sectors are not natural or merely technical, but are themselves socially and politically determined. The Rightward drift of politics in Britain affords some gross spectacles of what this can mean for two very different industries. On one side enormous sums of public money are poured into military activities – and thus into the sunrise electronics industry – with little real evaluation of the opportunity costs to the civilian economy in general. On the other, public investment for socially useful activities – such as low-tech construction work for house building – has been drastically reduced. Without a radical reappraisal of our dominant political priorities the former trend will continue to shackle our economic development, while the latter trend testifies to the poverty of progress in a supposedly high-tech capitalist society.

Notes

Chapter 1   Introduction

1   We shall use 'industry' and 'sector' interchangeably to avoid repetition.
2   Many developing countries have also started such programmes (Ernst, 1985).
3   We shall henceforth use the term 'radical' to cover a broad spectrum of views, from those of the British Labour Party through to the ultra-left.
4   Needless to say, there is not space to do justice to these methodological arguments here, but see Sayer (1984, especially ch. 4).

Chapter 2   Industry and Space

1   For a fuller discussion, see Sayer (1984).
2   See Marx (1973, p. 101), 'The concrete concept is concrete because it is a synthesis of many definitions, thus representing the unity of diverse aspects'.
3   See Burawoy (1979).
4   As Tomlinson argues, capital depends not on the existence of certain kinds of individuals as owners and controllers of capital but on the existence of a particular form of social relations and assorted practices, and these in turn may be supported by different forms of individuality without losing their character as capital, i.e. as institutions oriented towards expansion directly or indirectly through the production and sale of commodities for profit. Ownership and control are not properties of specific kinds of individuals as such but of particular social relations. Even though these relations only exist

where people reproduce them, they have powers irreducible to those individuals and, in the case of many modern firms, the corporation *per se* has effective ownership (Tomlinson, 1982).

5 See Marx (1976, pp. 475-7).

6 There are other qualifications which could be made but there is no need to refine the distinction further at this stage. However, see Walker (1985a).

7 See Murray (1972).

8 The neglect of product innovation probably arises because abstract theory in neoclassical and Marxist economics chooses to abstract from the disequilibrating and complicating effects of product innovation, preferring to consider economic processes with respect to a fixed set of types of use values (cf. Sayer, 1985). Taken to its logical conclusion, the abstraction implies a curious economy in which the value of commodities, and hence the possibility for making profits, is steadily driven downwards. Ignoring product innovation inevitably makes capital accumulation seem more crisis-prone than it really is, because (a) the resulting picture fails to distinguish what are in fact different markets and hence exaggerates the extent of price competition, and (b) it ignores the scope for finding new sources of surplus value.

9 '... in capitalist reality as distinguished from its textbook picture, it is not that kind of competition which counts (i.e. price competition) but the competition from the new commodity, the new technology, the new source of supply, the new type of organization ... competition which strikes not at the margins of the profits and outputs of the existing firms but at their foundations and their very lives' (Schumpeter, 1974, p. 84). Note that here, and in adjoining passages, Schumpeter does not deny the links between price and non-price competition.

10 See Harvey (1975) and Smith (1984).

11 In one respect figure 2.1 could be misleading, for there are also extensive organizational structures dealing with marketing of mass consumer products, but because of their routine nature these tend to be treated (actually or analytically) as separate and as part of the retail industry.

12 For a recent 'Bravermanian' paper see Locksley's (1986) article on information technology which interprets the 'purpose' of IT as primarily being the control of labour and which assumes (at least in its theoretical claims) that IT helps create more surplus only by process innovations.

13 It follows from this that, despite our interest in the labour process and its associated management–labour relations, we are more wary than many of using terms like 'Taylorism' and 'Fordism' as universal categories.

14 Markusen (1985) has recently developed a more sophisticated version of the model but it is vulnerable to the same criticisms.

15 See, for example, Markusen *et al.* (1986) and Langridge (1984).

16 See Markusen *et al.* (1986) and Kaldor *et al.* (1986) on the role of the defence industry within high tech.

17 See also Saxenian (1985).

18 Since this book was first drafted, Sinclair's firm has run into financial difficulties and his status as prophet of Britain's industrial renaissance has been quietly forgotten.
19 Our research methods are discussed more fully in Sayer (1982, 1984) and Morgan and Sayer (1985).

Chapter 3 Semiconductors

1 However, Greater Los Angeles has more jobs across a wider range of electronics.
2 Again, sources differ on such statistics; some put Western Europe's share of world demand at a quarter (Borrus, 1985; Malerba, 1985).
3 Truel, quoted in Scott (1986a).
4 See Grossman (1980); Global Electronics Newsletter (various dates); Elson and Pearson (1981); Fröbel et al. (1979); Lin (1985); Wong (1985).
5 This second factor works on the same principle as that underlying the 'just-in-time/total quality control' system pioneered in Japan in the automobile industry and now diffusing to other sectors and countries (Sayer, 1986).
6 In retrospect, Rada's claims concerning the return of assembly to the advanced countries seem to have been greatly exaggerated (United Nations Center on Transnational Corporations, 1982). Our thanks to Dieter Ernst for comments on this subject.
7 Recent political instability in the Philippines has discouraged inward investment there.
8 According to Paul (1985) government control over banks' investments has recently weakened so that they have become more market determined.
9 This is for permanent employment alone; in Toshiba, the six-month contracts of 2,000 temporary workers were not renewed.
10 See Cusumano (1985).
11 See Sayer (1986) for a discussion of these aspects in relation to the car industry.
12 Similarly, the head of Philips semiconductor planning said 'Europe simply does not provide the "market pull" needed to create world product standards' (de Jonquieres, 1985b).
13 It is thanks to the latter that Philips is the only European firm in the top ten merchant semiconductor companies.
14 Indeed, as the *Electronics Times* noted, one could virtually forecast such deals with the aid of a list of firms and a random number generator (11 March 1982, editorial, p. 10).
15 US complaints about Japan's predatory pricing policies in the memory chip market bear an uncanny resemblance to European complaints about US pricing policies for digital logic chips in the early 1970s. For example, the president of National Semiconductor, one of the

leading US merchant chip firms, freely admits that 'we went into the market with a vengeance. We were just plain heartless on price. We kept driving the price until one by one the other guys started leaving the game' (Malone, 1986).

16 The contortions that the merchant chip producer lobbyists went through is illustrated by the following statement by Robert Noyce, vice-chairman of Intel:

> The public policy task – to sustain US high technology leadership – is a formidable one indeed. The task is to retain the innate creativity and competitiveness of the individual firms based on free trade and free capital markets, while coping with foreign competition which has chosen to intervene in the market place to reshape the comparative advantage. (Semiconductor Industry Association, 1984)

Chapter 4 Consumer Electronics

1 Since this was written Thorn has pulled out of television manufacture, the last major producer to do so.
2 General Electric has now withdrawn completely and, like Thorn, has sold its consumer electronics business to Thomson of France.
3 As Storper and Walker (forthcoming) observe, this is an awkward modification of an already inept metaphor, but it highlights the transformation of the industry.
4 However, Philips has part ownership of Matsushita!
5 In the case of France, imports were restricted by the celebrated means of requiring them to be routed via a tiny customs post in Poitiers.

Chapter 5 Computer Systems

1 We use the term 'system integrators' in preference to the older, and thoroughly misleading, 'original equipment manufacturer'.
2 See Rosenberg (1982).
3 DEC, the largest minicomputer firm in the USA, found its shares halved in value when the new IBM PC was announced.
4 See Blackbourn (1975) for a discussion of the methods used by IBM in selecting locations. It is apparent from Blackbourn's account that political influence rates very high in IBM's assessments of investment possibilities.
5 Similarly, Hewlett-Packard, which makes mainly minicomputer systems, has no less than 240 sales offices in 75 countries and 19 manufacturing operations outside the USA. Over half of the latter include research laboratories, some of them developing products for world-wide distribution.

Chapter 6 Telecommunications

1   It is especially advanced in terms of the convergence of telecommunications and data processing; one index of this is that the USA has five times as many data terminals connected to its telephone network as the other OECD countries combined (Organization for Economic Cooperation and Development, 1983).

2   A classic illustration here being NEC's attempt to penetrate the US digital switch market in the 1970s with a product which was never destined for the Japanese market. Its effort failed largely because of intractable software problems and this, in turn, has been ascribed to NTT's relatively late move into digital switching for Japan's public network (Borrus et al., 1984).

Chapter 7 Electronics in Global Perspective

1   We do not have sectoral employment figures for any of the newly industrializing countries but presumably countries like South Korea have also experienced employment growth in this sector.

2   See chapter 2 for explanation of terms.

3   In effect we are arguing for a 'structurationist' approach *à la* Giddens (1979), as opposed to a structuralist approach.

4   For a critique of this work in the context of the semiconductor industry, see Sayer (1986).

5   One could of course argue that locations near customers are consistent with a cost minimization strategy, if one considers the costs of addressing them at a distance with equal effectiveness.

6   This problem is endemic in social scientific accounts of action (see Bourdieu, 1977).

7   While it is true to say that differences in national transmission standards were a restraining factor here, it must equally be noted that Japanese firms were less easily deterred by these problems in their export efforts.

Chapter 8 The Electronics Industry and Uneven Development in Britain

1   Similar relationships hold for equivalent data for electronic data-processing equipment (MLH 3302 from the new Standard Industrial Classification).

2   The company was denationalized in 1978.

3   The opportunity costs of this huge military complex are addressed in part IV.

4   There are many discussions of the typical 'IBM man', for instance; even Nancy Foy, in her book on IBM, hardly mentions women except as 'IBM wives' (Foy, 1974).

5   The reverse also is common; in one printed circuit board firm there was a switch of drilling jobs from women to men on the introduction of computer numerically controlled drills, changing the male–female ratio in the firm from 40:60 to 60:40. The manager explained this by the fact that drilling now involved some rudimentary engineering skills which women (allegedly) did not possess. Yet, when asked whether they had ever allowed women to try or considered training them, he admitted they had not (Source: interview).

6   While it is easy to explain the changes in the gender composition of the industry, it is far more difficult to explain the origin of the segregation and why electronics is more feminized than many other manufacturing industries. See Milkman (1983) and Walby (1987).

7   In 1981 the south-east accounted for 64.5 per cent of UK employment in electronic capital goods, 64 per cent in consumer electronics, 59 per cent in computers, 45.9 per cent in components, 34.4 per cent in telecommunications and overall for 53 per cent of UK employment (1981 Census of Employment).

8   That is, telecommunications, components, consumer, computer and electronic capital goods.

9   'Electronics' again refers to MLH 363–367, Census of Employment.

10 Variations in such circumstances account for differences in employment change; Greater Manchester, with its involvement in computer and capital equipment, had the second largest increase in electronics employment of any county in the period 1978–81.

11 This figure of 43,800 includes 1980 SIC activity headings 3710 and 3732 from instrument engineering, though these are not part of our definition of electronics in the rest of the book.

12 Indeed no less than eight instances of this were observed.

13 See figure 11.1.

14 Cf. Fothergill and Vincent (1985).

15 Here and henceforth, 'South Wales' refers to the counties of Gwent, South Glamorgan, West Glamorgan and Mid Glamorgan, i.e. *industrial* South Wales.

16 This category is wider than electronics but, given the size and character of the non-electronic part, we are confident this does not distort the comparison. Analysis of firm-specific data supports this.

17 Orwell called it 'probably the sleekest landscape in the world', quoted in Massey (1984, p. 142).

Chapter 9   A Modern Industry in a Mature Region

1   The Regional Employment Premium – this was a sexist policy for it subsidized male employment more than female (and juvenile) employment.

2   In-depth interviews were conducted with managers at all these plants, providing much of the information described in this chapter.

3   This was also true of the third Japanese television plant.

4   The first video recorder assembly was begun in 1986.
5   Recall the 'parts per million' quality standards described in chapter 4.
6   Compare the small new technology company which also located in England (chapter 10, p. 184).
7   By 1985-6, both Acorn and Sinclair had got into financial difficulties, the former being bailed out by the Italian computer firm Olivetti, the latter partly taken over by Amstrad.
8   See chapter 10.
9   The 'cultural' divide between such workers and the small number of blue collar workers in the prototype workshop was symbolized in the design of the site: the former were accommodated in a new high-tech building meant to impress customers as well as appeal to staff, the latter in some sheds at the back of the site separated from the former by a large car park. See Massey's (1984, pp. 147-9) observations on such divisions.

Chapter 10   Social Innovations 1

1   South Wales miners probably earned less than this average.
2   The Royal Commission on the Distribution of Income and Wealth considered workers receiving less than £85.5 per week to be low paid, and a family of two adults and two children would have needed £104.31 per week to have the same spending power as a similar family living on Supplementary Benefit (Low Pay Unit, 1983). As far as we know, foreign firms do not uniformly pay above local average rates and in the case of one Japanese firm operators were paid £60 for a 40-hour week while other Japanese plants were paying £80 for similar work. Generally, there was little evidence of clustering around a regional norm and pay rates differed significantly between plants.
3   General, Municipal, Boilermakers' and Allied Trades Union.
4   In Friedman's terms this marks a shift from 'direct control' to 'responsible autonomy' managerial strategies (Friedman, 1986).
5   Electrical, Electronic, Telecommunications and Plumbing Union.
6   Ironically, and in rather an un-Japanese fashion, this plant has now been sold off.
7   There are precedents for this managerial resistance; according to Littler (1982) and Urry (1986) managers were often the primary obstacle to the diffusion of 'scientific management' in the twenties and thirties.
8   Amalgamated Union of Engineering Workers.
9   A secondary consideration is that dealing with a single union simplifies negotiations; Takamiya contrasts the simple structure of industrial relations in one of the Japanese television plants in South Wales with those in a British-owned plant in England, which had to negotiate with seven different union teams, many of them operating in different configurations for different establishments within the same site (Takamiya, 1981).
10   Recall the Maynard survey categories, above.
11   For example, a survey of East Kilbride new town (only 9 miles south of Glasgow, a traditional bastion of organized labour) revealed

a high proportion of non-unionized plants – and not merely in small firms – in and beyond the electronics industry (East Kilbride Development Corporation, 1982).

12   We are grateful to Jo Foord for suggestions on this.

13   'Encouraged by the government the Old Man (IBM's boss) picked a location right in the heart of Scotland's "red Clydeside", one of the parts of the UK most troubled by industrial unrest. With skilled unemployment high, he reasoned that it is easier to have noticeable success if you start with abject failure. Yet Honeywell's factories at Newhouse, Bellshill and Uddington, in more serene parts of Scotland, suffered several months of strikes in the summer of 1972 . . .' (Foy, 1974, p. 169).

14   This aversion to the 'mass collective worker' was prevalent and is part of the emerging conventional wisdom among management.

15   Barr (1984). The prominence of the Japanese in automobiles and the fact that they pioneered the revolutionary 'just-in-time' method of production organization in automobiles accounts for the many examples of such changes in Western car firms; see Sayer (1986).

16   Personal communication, United Trades and Labor Council of South Australia, 11 South Terrace, Adelaide 5000.

17   See Cockburn (1983).

18   There are some positive signs on this front, however. Because of a fear that single-union, no-strike deals are a recipe for low unionization, trade unions in Wales and the North are cooperating in a way which could spell the end of some of the traditional conventions here. For example, they are proposing that these deals should be monitored and, if low unionization persists, other unions should be free to try to organize, rather than backing away as they are supposed to do at present.

Chapter 11   The Electronics Industry in the M4 Corridor

1   The former figure comes from Department of Employment data using a narrow definition of electronics, the latter from an estimate by Berkshire County Council for 1984 using a wider definition incorporating warehousing and software. Unfortunately no further details are given (Royal County of Berkshire, 1985). The difficulties of classification are all the greater where front-end activities such as those dominant in the corridor are abundant. Moreover, much of Avon's aerospace industry, perhaps 4,000 employers, is essentially doing electronics work. In the case of the Department of Employment data, further difficulties arise from reclassification between minimum list headings; the 1981 Berkshire figure is thought to be inflated roughly by 3,000 as a result of this.

2   The Berkshire 1984 survey found that 12 per cent of firms accounted for 66 per cent of employment (Royal County of Berkshire, 1985).

3   Some firms primarily concerned with software which would be unlikely to be classified in the 1968 Standard Industrial Classification under electronics were included, though plants or firms involved

primarily in warehousing and distribution were not included. 33 plants were surveyed. An unabbreviated version of what follows is available in the Library of the Economic and Social Research Council, London, entitled 'The performance of electronics companies in the M4 corridor', published in November 1986.

4   See chapter 2 for qualifications on the use of such terms.

5   The same firm was also given, as a gift, a semiconductor manufacturing shop located in north-west England.

6   This often involved using these other firms as system integrators or performing this role for them.

7   E's total employment world-wide stood at 1,784 in 1983 with 110 at its M4 site.

8   Total employment at the time of survey was 500.

9   H was set up in London in 1976 and moved out to the corridor in 1982 where it had 140 employees at the time of survey.

10   I has 200 employees in the corridor, J has 350.

11   Recall that the US market is over 20 times bigger than the UK one.

12   Six firms were visited in this group.

13   N had 380 employees at its M4 site.

14   This section is based on more cases than those summarized above.

15   The latter option was also found in two companies not mentioned above which were involved in making hybrid circuits (midway between printed circuit board and IC technologies).

16   This confirms Massey's conjecture that firms in such sectors and areas '. . . compete through technology and through quality, innovation and new products, at least as much as through price. What is being bought is not just labour power but scientific knowledge' (Massey, 1984, p. 142).

17   On subcontracting see Holmes (1986) and Scott (1983).

18   We are excluding from our definition 'spin-offs' of staff to other already-established firms; also cases of setting up of new branches of existing firms to service other firms.

19   An internal DTI survey found no evidence of personnel leaving (Department of Trade and Industry, 1982). Davies cites one case of a Newbury firm's founders coming from Harwell and one who invented the product which he later exploited while at a local university (Davies, 1984).

20   Nor should we assume that all start-ups are small and local: some may involve contact networks and mobilization of capital on a national, even international, scale, e.g. Inmos).

21   See Barlow and Savage (1986).

22   One of the main issues in planning for so-called high-tech industry is the fact that it is often impossible to distinguish from offices (Royal County of Berkshire, 1986).

23   In 1984, the average price of houses in Berkshire was £49,000, compared with £40,000 in the rest of the south-east and £31,000 nationally. In Bracknell, firms like 3M, Ferranti, Honeywell and Racal all offer large relocation packages to key recruits and Barlow and Savage estimate

that about 19 per cent of workers in Berkshire receive housing aid from their employers (Barlow, 1987). However, this applies only to a small minority of high-status workers, the ones least likely to be hampered by housing costs.

24   Our thanks to Sheila Honey for information on Bracknell.

Chapter 12   Social Innovations 2

1   See Cockburn (1983, p. 140).
2   Recall the quotation from Davies's (1984) survey in the previous chapter.
3   This middle-class equivalent of overtime has obvious implications for gender relations in the home.
4   See chapter 5.
5   See Friedman (1986); Lyn Eynon, personal communication.
6   This number of interviews was unique given the critical shortage of such skills.
7   In two of the M4 firms new practices which proved successful in front-end activities were later introduced into manufacturing at their Scottish plants.
8   According to one manager, who had moved from firm A to firm B, this was one of the major differences: A's workers tended to bury their mistakes, while B's workers were encouraged to reveal them.
9   To borrow Burns and Stalker's terms, the picture here is close to the 'mechanistic', as opposed to the 'organic' model approximated by some of the smaller leading firms (Burns and Stalker, 1962).
10   As with IBM, the new entrants had more success in increasing identity with the firm so that the 'them-and-us' attitude was often externalized, with the result that the 'them' became the competitor firms.
11   Ironically, large numbers of middle managers on salaries of £14,000–£20,000 belonged to one of the main unions in this firm, Administrative, Scientific, Technical and Managerial Staff (ASTMS).
12   In the 1930s IBM introduced salaries for workers instead of an hourly wage, together with a 'no-redundancy' policy (Drucker, 1979). Drucker sees this as of equal significance in the history of US industrial relations to Ford's $5 day. However, an internationally coordinated effort is now under way to unionize IBM, a project which the International Metalworkers' Federation (IMF) considers to be the 'labour equivalent of putting a man on the moon' (Bassett, 1987). The IMF believes that the time has never been better for this project, since IBM is trying to shed 10,000 jobs in the USA and 1,000 jobs in Europe – part of what IBM calls an 'early retirement' policy, but which the IMF sees as a partial abrogation of the no-redundancy strategy (see International Metalworkers' Federation, 1987).
13   Our thanks to Tony Fielding at Sussex for reminding us of this.
14   In the case of one firm (see chapter 11), the advantage of the M4's labour markets was the major factor for its eastward diversification across the Severn.

15 Our thanks to James Barlow and Mike Savage at Sussex for their comments on this section.

Chapter 13 'Sunrise' Industry, Innovation and the State

1 For an excellent discussion of the problematical nature of some of this capitalist state theory, see Urry (1981).
2 Official sponsorship of non-unionism is straining relations between the Thatcher Government and the EETPU, the government's favourite union. The EETPU is desperately trying to win recognition in a number of foreign electronics firms in Scotland. However, these firms have come to the UK partly because the government has advertised the UK's 'non-union facilities'.
3 The Thatcher Government has never been well disposed to the tri-partite NEDC because of its general distaste for tri-partitism of any kind. Therefore it came as no surprise when, in July 1987, the government decided to abolish nearly half of the Economic Development Committees (EDCs) including the Information Technology Economic Development Committee (ITEDC), a body which had consistently warned of a coming crisis in the UK IT industry.
4 The IT 86 Committee's hopes for a major post-Alvey support programme for IT have already been dashed by the government's decision that private capital will have to finance the bulk of the costs of such a programme.

Chapter 14 Conclusions

1 Japanese mass production firms have now shown that, when coordinated on a just-in-time delivery basis, highly localized clusters of suppliers round final assembly plants can give rise to major productivity gains. Western firms are now beginning to imitate this; for example, IBM has encouraged several major suppliers to relocate close by its Montpellier (France) and Boca Raton (Florida, USA) plants in order to coordinate production more tightly (Bakis, 1980; *Business Week*, 1983; Sayer, 1986).
2 See *Transnational Information Exchange* (various dates).
3 Workerist accounts tend to make a double reduction – of human agency to workers' agency and of workers' agency to class struggle.
4 If it only involved the latter it would not be appropriate to call it 'contradictory'.
5 While there are links between needs, demand and profitability they are highly imperfect and in no way adequately resolve producer and consumer interests.
6 These combinations of interests, some conflicting, some common, lend themselves to game theory analysis; indeed there is now an emerging 'game-theoretic Marxism', e.g. Przeworski (1980) Offe and Wiesenthal (1980) and Lash and Urry (1984).

7 Too often assessments of the employment effects of new technology focus on only a few sectors and take little account of second-order economic effects.

8 This may be contributing to the widely noted tendency in recent years for vertical disintegration to increase.

9 The need for management skills has again been one of the main lessons of the Greater London Enterprise Board.

10 40 per cent of the UK market for IT products is in the public sector. Ieuan Maddock recommended a Civil Purchasing Policy Unit that would harness and publicize a new civil market. The 'relative lack of state investment in public administration systems . . . has placed UK firms at a disadvantage the larger "bespoke" systems houses in France [*sic*], where the level of statement investment in computerisation is much higher'. (Greater London Council, 1985, p. 324).

11 The trials and tribulations of the French socialist government provide instructive lessons in what can happen to uncritical commitments to nationalization: an indebted French government reaped ownership without control!

12 There need be no contradiction here between restricting inward investment and our previous comment on management–labour relations and job security in foreign firms, for the latter features could be achieved within British firms.

13 In the case of (c), it needs to be borne in mind that imports can also have labour displacement effects.

Bibliography

Abernathy, W. J., Clark, K. B. and Kantrow, A. M. (1984), *Industrial Renaissance: Producing a Competitive Future for America*, New York, Basic Books.

Advisory Council for Applied Research and Development (1978), *Industrial Innovation*, London, HMSO.

Advisory Council for Applied Research and Development (1980), *R&D for Public Purchasing*, London, HMSO.

Advisory Council for Applied Research and Development (1986), *Software: A Vital Key to UK Competitiveness*, London, HMSO.

Aglietta, M. (1979), *A Theory of Capitalist Regulation*, London, New Left Books.

Aoki, M. (1987), 'Why and how is inter-firm, inter-disciplinary R&D co-operation developing in Japan?', Mimeograph, Stanford University.

Armstrong, P., Glyn, A. and Harrison, J. (1984), *Capitalism Since World War 2: The Making and Break-up of the Great Boom*, London, Fontana.

Arnold, Erik (1984), *Competition and Technical Change in the UK Television Industry*, London, Macmillan.

Arnold, Erik and Guy, Ken (1986), *Parallel Convergence: National Strategies in Information Technology*, London, Frances Pinter.

Atkinson, John (1984), *Manning for Uncertainty: Some Emerging UK Work Patterns*, University of Sussex, Institute of Manpower Studies.

Bakis, H. (1980), 'A case study of IBM's global data network', paper presented to the 24th International Geographical Congress, Tokyo, August.

Bank of England (1985), 'Services in the UK economy', *Quarterly Review*, September.

Barlow, James and Savage, Mike (1986), 'Economic restructuring and housing provision in Berkshire', Urban and Regional Studies Working Paper 55, University of Sussex.

Barlow, James (1987), 'The housing crisis and its local dimensions', *Housing Studies*, forthcoming.

Barr, A. (1984), Address to the *Financial Times* Conference on Automated Manufacturing, quoted in the *Financial Times*, 29th March.

Bassett, P. (1986), *Strike Free: New Industrial Relations in Britain*, London, Macmillan.

Bassett, P. (1987), 'IBM faces pressure for union recognition', *Financial Times*, 13 January.

Bernstein, B., DeGrasse, B., Grossman, R., Paine, C. and Siegel, L. (1977), *Silicon Valley: Paradise or Paradox?*, Mountain View, CA, Pacific Studies Center.

Blackbourn, A. (1975), 'The spatial behaviour of American firms on Western Europe', in F. E. Ian Hamilton (ed.) *Spatial Perspectives on Industrial Organization and Decision-making*, London, Wiley.

Bloom, B. (1985), 'Windfall defence profits "should have been repaid"', *Financial Times*, 25 July.

Bloom, B. (1986), 'Defence expenditure under attack', *Financial Times*, 28 April.

Booz, Allen and Hamilton (1979), *The Electronics Industry in Scotland: Report to the Scottish Development Agency*, Glasgow, SDA.

Borrus, Michael (1985), *Reversing Attrition: A Strategic Response to the Erosion of US Leadership in Microelectronics*, Berkeley, CA, Berkeley Roundtable on the International Economy.

Borrus, Michael, Bar, F. and Cogez, P. (1984), *Telecommunications Development in Comparative Perspective: Report to the Office of Technology Assessment*, Berkeley, CA, Berkeley Roundtable on the International Economy.

Bourdieu, P. (1977), *Towards a Theory of Practice*, Cambridge, Cambridge University Press.

Braverman, Harry (1974), *Labor and Monopoly Capital*, New York, Monthly Review Press.

Brittan, S. (1985), 'Coronets and begging bowls', *Financial Times*, 17 October.

Brock, G. (1981), *The Telecommunications Industry: The Dynamics of Market Structure*, Cambridge, MA, Harvard University Press.

Brown, Gordon (ed.) (1975), *The Red Paper on Scotland*, Edinburgh, EUSPB.

Brown, W. (1983), 'The emergence of enterprise unionism', Address to the Institute of Personnel Management, Harrogate.

Buiter, Wilhelm and Miller, Max (1983), 'Changing the rules: economic consequences of the Thatcher regime', *Brookings Papers on Economic Activity*, 2, 305–79.

Burawoy, Michael (1979), *Manufacturing Consent*, Chicago, University of Chicago Press.

Burawoy, Michael (1983), *The Politics of Production*, London, Verso.

Burns, T. and Stalker, G. M. (1962), *The Management of Innovation*,

London, Tavistock.

Burton, J. (1979), *The Job Support Machine: A Critique of the Subsidy Morass*, London, Centre for Policy Studies.

Business Week (1983), 'How the PC project changed the way IBM thinks', 3 October.

Business Week (1984), 'Reshaping the computer industry', 16 July.

Business Week (1985a), 'Labor relations: we have the right mix', 28 January.

Business Week (1985b), 'America's high-tech crisis: why Silicon Valley is losing its edge', 11 March.

Business Week (1986), 'National Semiconductor wants to change its spots', 28 April.

Butler, S. B. (1985), 'Fishing for big gains in chips', *Financial Times*, 9 May.

Cabinet Office (1986), *Annual Review of Government Funded R&D*, London, HMSO.

Cable, V. and Clarke, J. (1981), *British Electronics and Competition with Newly Industrialising Countries*, London, Overseas Development Corporation.

Cambridge Econometrics (1985), *Electrical Engineering Industries*, Cambridge, Cambridge Econometrics.

Carpentier, M. (1985), Address to the *Financial Times* Conference on World Telecommunications, London, December.

Cawson, Alan, Holmes, Peter, Morgan, Kevin, Stevens, Ann and Webber, Doug (1988), *Unsteady State: Government–Industry Relations in European Electronics*, Oxford, Oxford University Press.

Chang, Y. S. (1971), 'The transfer of technology: the economics of offshore assembly: the case of the semiconductor industry', *United Nations Institute for Training and Research*, 11, New York, UNITAR.

Commission of the European Communities (1984a), *Second Periodic Report on the Regions of the Community*, Brussels, CEC.

Commission of the European Communities (1984b), *Communication from the Commission to the Council on Telecommunications*, Brussels, CEC.

Cockburn, Cynthia (1983), *Brothers*, London, Pluto.

CSE Microelectronics Group (1980), *Microelectronics: Capitalist Technology and the Working Class*, London, CSE Books.

Cooke, Philip (1981), 'Tertiarisation and socio-spatial differentiation in Wales', *Geoforum*, 12, 319–30.

Cooke, Philip (1984), 'Region, class and gender: a European comparison', *Progress in Planning*, 22(2), 85–146.

Coopers and Lybrand (1987), *Computing Services Industry 1986–96: A Decade of Opportunity*, London, Coopers and Lybrand.

Crewe, Ivor (1987), 'A new class of politics', *Guardian*, 15 June.

Critchley, Julian (1987), 'No sign of a regional policy', *Guardian*, 27 February.

Crisp, Jason (1984), 'Japan keeps up the pressure', *Financial Times*, 28 March.

Cusumano, M. A. (1985), *The Japanese Automobile Industry*, Cambridge, MA, Harvard University Press.

Davies, Andy (1984), 'The electronics industry in West Berkshire: a case study', Mimeograph, University of Sussex.

de Jonquieres, Guy (1983), 'The battle of the two corporate giants', *Financial Times*, 11 April.

de Jonquieres, Guy (1985a), 'Electronics trade deficit will deepen', *Financial Times*, 18 February.

de Jonquieres, Guy (1985b), 'Seeking to shift the emphasis on high technology grants', *Financial Times*, 26 March.

de Jonquieres, Guy (1985c), 'The harsh imperatives of survival', *Financial Times*, 24 June.

de Jonquieres, Guy (1985d), 'Why Europe is so rattled', *Financial Times*, 7 May.

de Jonquieres, Guy (1985e), 'Boiling oil for the chipmakers', *Financial Times*, 8 July.

Department of Industry (1982a), *A Programme for Advanced Technology: the Report of the Alvey Committee*, London, HMSO.

Department of Industry (1982b), *The Location, Mobility and Financing of the Computer Services Sector in the UK*, London, South East Regional Office.

Department of Industry (1983), *The Department of Industry's Strategic Aims*, NEDC (83)13.

Department of Trade and Industry (1982), *The Location, Mobility and Finance of New High Technology Companies in the U.K. Electronics Industry*, unpublished report, prepared by J. Beaumont.

Department of Trade and Industry (1983a), *Industrial Policy*, NEDC (83)49.

Department of Trade and Industry (1983b), *Regional Industrial Policy*, London, HMSO.

Department of Trade and Industry (1984a), *The Government View: Response to the IT–EDC Report*, London, DTI.

Department of Trade and Industry (1984b), *Inward Investment Review*, London, DTI.

Department of Trade and Industry (1985), *Science and Technology Report*, London, DTI.

Dicken, Peter (1986), *Global Shift*, London, Harper and Row.

Dickson, T. (ed.) (1980), *Scottish Capitalism: Class, State and Nation from Before the Union to the Present*, London, Lawrence and Wishart.

Dore, Ron P. (1985), 'Financial structures and the long term view', *Policy Studies*, 6(1), 10–29.

Dosi, Giovanni (1981), 'Technical change and survival: Europe's semiconductor industry', *Industrial Adjustment and Policy II*, Sussex European Research Centre Paper 9, University of Sussex.

Dosi, Giovanni (1982), 'Technical change and industrial transformation – the theory and an application to the semiconductor industry', Science Policy Research Unit, University of Sussex, pp. 323–4.

Drucker, Peter (1979), *Management*, London, Pan.

Duncan, Mike (1981), 'The information technology industry in 1981', *Capital and Class*, 17, 79–113.

East Kilbride Development Corporation (1982), *East Kilbride: a Labour Study*, East Kilbride, EKDC.

Economist Intelligence Unit (1985), *The UK as a Tax Haven*, London, EIU.

Edwards, Nicholas (1984), 'Industrial change and high technology', *Agenda*, 5(summer), 3–9.

Electronic Consumer Goods Sector Working Party (1983), *Evidence to House of Lords Select Committee on Science and Technology, Engineering Research and Development*, London, HMSO.

Electronics EDC, (1982), *Policy for the UK Electronics Industry*, London, NEDO.

Electronics Capital Equipment EDC (1983), *The Future for the Electronic Capital Equipment Industry: In Whose Hands?*, London, NEDO.

Electronics Industrial Association (1986), *Market Data Research Book*, London, EIA.

The Engineer (1984), 'The taming of the unions', 15 March.

Elson, Diane and Pearson, Ruth (1981), 'Nimble fingers make cheap workers: an analysis of women's employment in Third World export manufacturing', *Feminist Review*, 7, 87–107.

Engineering Employers' Federation (1987), *Towards an Industrial Strategy*, London, EEF.

Engineering Industry Training Board (1982), *Manpower and Training in the Electronics Industry*, Watford, EITB.

Engineering Industry Training Board (1985), personal communication.

Ergas, Henry (1985), 'Exploding the myths about what's wrong', *Financial Times*, 26 June.

Ergas, Henry (1986), *Does Technology Policy Matter?*, Paris, OECD.

Ernst, Dieter (1981), *Restructuring World Industry in a Period of Crisis: the Role of Innovation*, New York, UNIDO.

Ernst, Dieter (1985), 'Automation and worldwide restructuring of the electronics industry', *World Development*, 13(3), 333–52.

Ernst, Dieter (1986), 'U.S.–Japanese competition and the worldwide restructuring of the electronics industry – a European view', paper prepared for the Pacific–Atlantic Interrelations Conference, Institute of International Relations, University of California, Berkeley, 24 April.

Eurogestion (1986), *Trends in Information Technology*, Brussels, Eurogestion.

European Research Associates (1986, *Inter-EEC and Extra-EEC Trade Flows in Telecommunications Equipment*, Brussels, ERA.

Financial Times (1986), 'Industry's besetting sin', 5 September.

Firn, John and Roberts, David (1984), 'High technology industries', in N. Hood and S. Young (eds) *Industrial Policy and the Scottish Economy*, Edinburgh, Edinburgh University Press.

Fothergill, Stephen and Gudgin, Graham (1982), *Unequal Growth*,

London, Heinemann.

Fothergill, Stephen and Vincent, Jill (1985), *The State of Britain*, London, Pan.

Foy, Nancy (1974), *The IBM World*, London, Eyre Methuen.

Freeman, Christopher (1979), 'Technical innovation and British trade performance', in F. Blackaby (ed.) *De-industrialisation*, London, Heinemann.

Freeman, Christopher (1987), *Technology Policy and Economic Performance: Lessons from Japan*, London, Pinter.

Friedman, Andrew (1986), 'Developing the managerial strategies approach to the labour process', *Capital and Class*, 30, 97–124.

Friedman, Andrew and Cornford, Dominic (1985), 'The future for programmers: analyst programmers or mere coders?', Mimeograph, University of Bristol.

Fröbel, Fokker, Heinrichs, Jurgen and Kreye, Otto (1979), *The New International Division of Labor*, Cambridge, Cambridge University Press.

Gardiner, Paul and Rothwell, Roy (1985), *Innovation: A Study of the Problems and Benefits of Product Innovation*, London, Design Council.

General Electric Company (1985), *Report and Accounts*, London.

General, Municipal, Boilermakers and Allied Trades Union (1981), 'Model agreement', Mimeograph, South Wales Region GMBATU, Cardiff.

Giddens, Anthony (1979), *Central Problems of Social Theory*, London, Macmillan.

Gillan, W. (1986), *The Japanese Secret: Are They Winning?*, London, Department of Trade and Industry.

Glasmeier, Amy (1986), 'Factors governing the development of high technology complexes', paper presented to the Anglo-American Symposium on the Location of High Technology Industry, University of Cambridge, July.

Global Electronics Newsletter, various dates.

Godley, Wyn (1987), 'A manifesto that plays with mirrors', *Observer*, 7 June.

Gordon, R. and Kimball, L. (1986), 'Industrial structure and the changing global dynamics of location in high technology industry', Silicon Valley Research Group Working Paper 3, University of California, Santa Cruz.

Greater London Council (1985), *The London Industrial Strategy*, London, GLC.

Greater London Council (1986), *The London Technology Strategy*, London, GLC.

Gregory, Derek and Urry, John (1985), *Social Relations and Spatial Structures*, London, Macmillan.

Grossman, Rachel (1980), 'Bitter wages: women in East Asia's semiconductor plants', *Multinational Monitor*, 1(2), 8–11.

Hall, Peter, Breheny, Michael, McQuaid, Ronald and Hart, Douglas (1987), *Western Sunrise*, London, Allen and Unwin.

Hall, Stuart (1985), 'Authoritarian populism: a reply to Jessop et al.', *New Left Review*, 15, 115–23.

Harris, Frank and McArthur, Richard (1985), 'The issue of high technology: an alternative view', Working Paper 16, North West Industry Unit, School of Geography, University of Manchester.

Harvey, David (1975), 'The geography of capital accumulation', *Antipode*, 7(2), 9–21.

Harvey, David (1982), *The Limits to Capital*, Oxford, Blackwell.

Henderson, Jeff (1985), 'The new international division of labour and American semiconductor production in South-East Asia', in D. Watts, C. Dixon and D. Drakakis-Smith (eds) *Multinational Companies and the Third World*, London, Croom Helm.

Henderson, Jeff (1986), 'Semiconductors, Scotland and the international division of labour', Mimeograph, Centre of Urban Studies and Urban Planning, University of Hong Kong.

Hills, Jill (1985), 'Industrial policy and the information technology sector', Mimeograph, Greater London Council.

Hills, Jill (1986), *Deregulating Telecoms*, London, Frances Pinter.

Holmes, J. (1986), 'The organization and locational structure of production subcontracting', in A. J. Scott and M. Storper (eds) *Production, Work, Territory: the Geographical Anatomy of Industrial Capitalism*, London, Allen and Unwin.

Hood, N. and Young, S. (1983), *Multinational Investment Strategies in the British Isles: Report to the Department of Trade and Industry*, London, HMSO.

House of Lords (1983), *Select Committee on Science and Technology, Engineering Research and Development*, Volumes I–III, London, HMSO.

House of Lords (1985), *Select Committee on Overseas Trade*, Volumes I–III, London, HMSO.

House of Lords (1986), *Select Committee on Science and Technology, Civil Research and Development*, Volumes I–III, London, HMSO.

Hymer, Stephen H. (1972), 'The multinational corporation and the law of uneven development', in J. N. Bhagwati (ed.) *Economics and World Order*, New York, Macmillan.

IBM UK Ltd (1984), Company Report, Havant, IBM.

IBM UK Ltd (1985), *Evidence to House of Lords Select Committee on Overseas Trade*, 30 April.

IBM UK Ltd (1986), *Memorandum to House of Lords Select Committee on Science and Technology, Civil Research and Development*, Vol. III, 14 June, 137–40.

IEEE Spectrum (1986), 'Assessing Japan's role in telecommunications', June, 47–52.

Ikeda, M. (1979), 'The subcontracting system in the Japanese electronics industry', *Engineering Industries of Japan*, 19, 43–71.

Income Data Services (1984), *Group Working and Greenfield Sites*, Study 314, London, IDS Ltd.

IT 86 Committee (1986), *Information Technology: A Plan For Concerted Action*, London, HMSO.

Information Technology EDC (1984), *Crisis Facing UK Information Technology*, London, NEDO.

International Metalworkers' Federation (1987), *The IBM File*, Geneva, IMF.

Invest in Britain Bureau (1984), *Annual Report*, London, Department of Trade and Industry.

Invest in Britain Bureau (1983), *Inward Investment and the IBB: 1977–1982*, London, Department of Trade and Industry.

Irwin, Manley (1984), *Telecommunications America: Markets Without Boundaries*, New York, Quorum Books.

Jay, A. (1972), *Corporation Man*, London, Cape.

Joseph, Keith (1981), 'Regional policy', *Hansard*, 8 July, cos 427–33.

Kaldor, Mary (1982), *The Baroque Arsenal*, London, Andre Deutsch.

Kaldor, Mary, Sharp, Margaret and Walker, William (1986), 'Industrial competitiveness and Britain's defence commitments', *Lloyds Bank Review*, October.

Keeble, David (1976), *Industrial Location and Planning in the UK*, London, Methuen.

Kehoe, L. (1984), 'Bid to regain world leadership for US in chip technology', *Financial Times*, 14 May.

Kehoe, L. (1986), 'US suffers $8.8bn deficit in electronics', *Financial Times*, 15 August.

Kelly, John (1982), *Scientific Management, Job Redesign and Work Performance*, London, Academic Press.

Labour Research (1987), 'UK firms seek rosier climes', May.

Langridge, B. (1984), 'Defining high technology industries', Department of Economics Working Paper, University of Reading.

Large, Peter (1985), 'Is Britain bound for the sweaty league?', *Guardian*, 27 March.

Lash, Scott and Urry, John (1984), 'The new marxism of collective action: a critical analysis', *Sociology*, 18(1), 33–50.

Lee, S. M. and Schwendiman L. (eds) (1982), *Management by Japanese Systems*, New York, Praeger.

Lin, L. (1985), 'Health, women's work, and industrialization: women workers in the semiconductor industry in Singapore and Malaysia', paper presented to the International Sociological Association/Center of Urban Studies and Urban Planning Conference, University of Hong Kong, 14–20, August.

Lipietz, A. (1984), 'Imperialism or the beast of the apocalypse', *Capital and Class*, 22, 81–109.

Lipsey, David (1985), 'Will industry learn the Lawson?', *Sunday Times*, 17 November.

Littler, Craig (1982), *The Development of the Labour Process in Capitalist Societies*, London, Heinemann.

Littler, Craig and Salaman, Graham (1984), *Class at Work*, London, Batsford.

Locksley, Gareth (1986), 'Information technology and capitalist development', *Capital and Class*, 27, 81–106.

Lovering, John (1985), 'Defence expenditure and the regions: the case of Bristol', *Built Environment*, 11(3), 193–206.

Lovering, John (1986), 'The restructuring of the defence industries', School for Advanced Urban Studies Working Paper 59, University of Bristol.

Low Pay Unit (1983), 'The case for a national minimum wage', Pamphlet no. 23, London, Maynard & Co.

Ludolph, C. M. (1984), *Trends in High Technology, Trade and Investment*, New York, U.S. Department of Commerce, International Trade Administration.

Mackenzie, Ian W. (1985), 'Advanced computing: the issues for Europe', Mimeograph, Centre for Business Strategy, London Business School.

Mackintosh, Ian (1986), *Sunrise Europe: The Dynamics of Information Technology*, Oxford, Blackwell.

McEwan, A. (1985), 'Unstable empire: US business in the international economy', *IDS Bulletin*, 16(1), 40–6.

McLennan, S. (1984), *The Coming Computer Industry Shakeout*, Chichester, Wiley.

McKinsey & Co. (1983), *A Call to Action: The European IT Industry*, Brussels, McKinsey & Co.

Maddock, I. (1982), *Evidence to House of Lords Select Committee on Science and Technology, Engineering Research and Development*, Vol. III, 24 June, 375–90.

Maddock, I. (1983), *Civil Exploitation of Defence Technology*, London, NEDO.

Malerba, F. (1985), 'Demand structure and technological change: the case of the European semiconductor industry', *Research Policy*, 14, 283–97.

Malone, M. (1986), *The Big Score: The Billion Dollar History of Silicon Valley*, New York, Doubleday.

Manpower Services Commission (1984), *Competence and Competition: Training and Education in the FRG, US and Japan*, London, NEDO.

Marglin, Stephen (1976), 'What do bosses do?', in Andre Gorz (ed.) *The Division of Labour*, Hassocks, Harvester.

Markusen, Ann Roell (1985), *Profit Cycles, Oligopoly and Regional Development*, Cambridge, MA, MIT Press.

Markusen, Ann Roell, Hall, Peter and Glasmeier, Amy (1986), *High Tech America*, Boston, Allen and Unwin.

Marsh, Peter (1985), 'Britain's R&D programme: a disturbing outlook', *Financial Times*, 3 December.

Marx, Karl (1973), *Grundrisse*, Harmondsworth, Penguin.

Marx, Karl (1976), *Capital*, Volume 1, Harmondsworth, Penguin.

Massey, Doreen (1979), 'In what sense a regional problem?', *Regional Studies*, 13, 233–43.

Massey, Doreen (1984), *Spatial Divisions of Labour*, London, Macmillan.

Massey, Doreen and Allen, John (eds) (1985), *Geography Matters!*, Cambridge, Cambridge University Press.

Massey, Doreen and Meegan, Richard A. (1979), 'The geography of industrial reorganization', *Progress in Planning*, 10(3), 155–237.

Maynard, H. B. & Co. (1978), *A Survey of Operating Conditions in Europe Experienced by US Owned Companies*, London, Maynard & Co.

Midland Bank Review (1986), 'UK manufacturing: output and trade performance', Autumn.

Milkman, Ruth (1983), 'Female factory labor and industrial structure: control and conflict over "Woman's Place" in auto and electrical manufacturing', *Politics and Society*, 12(2), 159–204.

Ministry of Defence (1986a), *Statement on the Defence Estimates*, 1, London, HMSO.

Ministry of Defence (1986b), *Statement on the Defence Estimates*, 2, London, HMSO.

Monopolies and Mergers Commission (1986), *British Telecommunications plc and Mitel Corporation: A Report on the Proposed Merger*, Cmnd. 9715, London, HMSO.

Morgan, Kevin (1983a), 'Restructuring steel: the crises of labour and locality in Britain', *International Journal of Urban and Regional Research*, 7(2), 175–201.

Morgan, Kevin (1983b), 'The politics of industrial innovation in Britain', *Government and Opposition*, 18(3), 304–18.

Morgan, Kevin (1985), 'Regional regeneration in Britain: the "territorial imperative" and the Conservative state', *Political Studies*, 4, 560–77.

Morgan, Kevin (1986), 'Reindustrialization in peripheral Britain: state policy, the space economy and industrial innovation', in Ronald Martin and Robert Rowthorn (eds) *The Geography of Deindustrialization*, London, Macmillan.

Morgan, Kevin (1987), 'Breaching the monopoly: telecommunications and the state in Britain', Working Paper Series on Government–Industry Relations No. 7, University of Sussex.

Morgan, Kevin (1988), 'High technology and regional development: for Wales, see Greater Boston?', in G. Day and G. Rees (eds) *Contemporary Wales*, Vol. I, Cardiff, University of Wales Press.

Morgan, Kevin and Sayer, Andrew (1985), 'A modern industry in a declining region: links between theory, method and policy', in D. Massey and R. A. Meegan (eds) *The Politics of Method*, London, Methuen.

Morgan, Kevin and Webber, Douglas (1986), 'Divergent paths: political strategies for telecommunications in Britain, France and the Federal Republic of Germany', *West European Politics*, 9(4), 56–79.

Murray, Robin (1972), 'Underdevelopment, the international firm and the international division of labour', in *Towards a New World Economy*, papers and proceedings of the Society for International Development, The Hague, 1971, Rotterdam, Rotterdam University Press.

National Economic Development Committee, Electronic Consumer

Goods Sector Working Party (1980), *Evidence to House of Lords Select Committee on Science and Technology Engineering R&D*, 24 February.

National Economic Development Committee (1983a), *Crisis in the IT Industry*, London, NEDO.

National Economic Development Committee (1983b), *Report on the Sector Assessments*, London, NEDO.

National Economic Development Office (1985), *Evidence to the House of Lords Committee on Overseas Trade*, London, HMSO.

National Economic Development Office (1987), *The Making of Managers*, London, NEDO.

National Institute of Economic and Social Research (1985), *Economic Review*, 11, London, NIESR.

Nayyar, D. (1979), 'Transnational corporations and manufactured exports from poor countries', *Economic Journal*, 88, 59–84.

Nelson, R. (ed.) (1982), *Government Support of Technical Progress: a Cross Industry Analysis*, New York, Pergamon.

Nelson, R. and Winter, S. (1977), 'Simulation of Schumpeterian competition', *American Economic Review: Papers and Proceedings*, 67, 271–6.

New, Colin (1986), 'Short-term gains and long-term disaster', *Sunday Times*, 15 June.

Newman, Rhona and Newman, Julian (1985) 'Information work: the new divorce?', *British Journal of Sociology*, 36(4), 497–515.

Offe, Claus and Wiesenthal, H. (1980), 'Two logics of collective action: theoretical notes on social class and organizational form', *Political Power and Social Theory*, 1, 67–115.

Office of Technology Assessment (1986), 'Plant closings: advance notice and rapid response', Washington, DC, Congress of the U.S.

Okimoto, D. and Hayase, H. (1985), 'Organisation for innovation: NTT laboratories and NTT family firms', Mimeograph, Stanford University.

Organization for Economic Cooperation and Development (1983), *Telecommunications: Pressures and Policies for Change*, Paris, OECD.

Organization for Economic Cooperation and Development (1984), *Foreign Trade Statistics*, Paris, OECD.

Organization for Economic Cooperation and Development (1985), *The Semiconductor Industry*, Paris, OECD.

Paul, J. K. (ed.) (1985), *High Technology, International Trade and Competition*, Park Ridge, NJ, Noyes Publications.

Pavitt, Keith (ed.) (1980), *Technical Innovation and British Economic Performance*, London, Macmillan.

Pike, Alan (1986), 'MSC chief warns on skill crisis', *Financial Times*, 20 May.

Posa, J. G. (1981), '"No hands" assembly packages chips', *Electronics*, 2 June.

Przeworski, A. (1980), 'Material bases of consent: economics and politics in a hegemonic system', *Political Power and Social Theory*, 1, 21–66.

Rada, Juan (1982), *Structure and Behaviour of the Semiconductor Industry*,

New York, UNCTC.

Rada, Juan (1985), 'New trends in the semiconductor industry', *Electronics Location File*, 7–10.

Radice, Hugo (1984), 'The national economy: a Keynesian myth?', *Capital and Class*, 22, 111–40.

Rapoport, C. (1986), 'The tiny challenger', *Financial Times*, 21 August.

Rentoul, John (1987), *The Rich Get Richer: The Growth of Inequality in Britain in the 1980's*, London, Unwin.

Rosenberg, N. (1982), *Inside the Black Box*, Cambridge, Cambridge University Press.

Rosenbloom, R. S. and Abernathy W. J. (1982), 'The climate for innovation in industry: the role of management attitudes and practices in consumer electronics', *Research Policy*, 11(4), 209–25.

Rothwell, Roy (1982), 'The role of technology in industrial change: implications for regional policy', *Regional Studies*, 16, 361–70.

Royal County of Berkshire (1985), *Report of the 1984 Survey of employers*, Reading, Berkshire County Council Planning Department.

Royal County of Berkshire (1986), *The Development of High Technology Industries in Berkshire*, Reading, Berkshire County Council Planning Department.

Saxenian, Anna-Lee (1984), 'The urban contradictions of Silicon Valley', *International Journal of Urban and Regional Research*, 7(2), 237–62.

Saxenian, Anna-Lee (1985), 'The genesis of Silicon Valley', in Peter Hall and Ann Markusen (eds) *Silicon Landscapes*, London, Allen and Unwin.

Sayer, Andrew (1982), 'Explanation in economic geography', *Progress in Human Geography*, 6, 6–15.

Sayer, Andrew (1983), 'Theoretical problems in the analysis of technological change and regional development', in F. E. I. Hamilton and G. L. R. Linge (eds) *Spatial Analysis, Industry and the Industrial Environment*, Vol. 3, Chichester, Wiley.

Sayer, Andrew (1984), *Method in Social Science: a Realist Approach*, London, Hutchinson.

Sayer, Andrew (1985), 'Industry and space', *Environment and Planning D: Society and Space*, 3, 3–29.

Sayer, Andrew (1986), 'Industrial location on a world scale: the case of semiconductors', in Allen J. Scott and Michael Storper (eds) *Production, Work, Territory*, London, Allen and Unwin.

Schiller, Dan (1982), *Telematics and Government*, Norwood, NJ, Ablex.

Schiller, Dan (1983), 'The storming of the PTTs', *Datamation*, May, 155–8.

Schumpeter, Joseph A. (1974), *Capitalism, Socialism and Democracy*, London, Unwin.

Sciberras, Ed (1980), *Study of Direct Investment in the UK by Japanese Enterprises*, University of Sussex, Science Policy Research Unit.

Science Policy Research Unit (1972), *Success and Failure in Industrial Innovation: Report on Project Sappho*, London, Centre for Studies of Industrial Innovation.

Scott, Allen J. (1983), 'Industrial organization and the logic of intra-metropolitan location, I: theoretical considerations', *Economic Geography*, 59, 223–50.

Scott, Allen J. (1986a), *The Semiconductor Industry in South-East Asia: organization, location and the international division of labour*, Mimeograph, Department of Geography, University of California, Los Angeles.

Scott, Allen J. (1986b), 'High technology industry and territorial development: the rise of the Orange County complex, 1955–1984', *Urban Geography*, 7, 3–45.

Scott, Allen J. and Angel, David P. (1986), *The US Semiconductor Industry: a locational analysis*, Mimeograph, Department of Geography, University of California, Los Angeles.

Scottish Development Agency (1982), *Electronics in Scotland: The Leading Edge*, Glasgow, SDA.

Semiconductor Industry Association (1984), *1983–84 Yearbook and Directory*, Palo Alto, CA, SIA.

Senker, P., Davies, A. and Lowe, S. (1985), *Information Technology and De-centralisation: Report to the EEC*, University of Sussex, Science Policy Research Unit.

Short, John (1981), 'Defence spending in the UK regions', *Regional Studies*, 15, 101–10.

Siegel, Lenny (1980), 'Delicate bonds: the global semiconductor industry', *Pacific Research*, 9(1), Special issue, 1–26.

Smith, David (1987), *The Rise and Fall of Monetarism*, London, Pelican.

Smith, M. (1985), 'Matsushita uproots for fresher electronics fields', *Financial Times*, 10 July.

Smith, Neil (1984), *Uneven Development: Nature, Capital and the Production of Space*, Oxford, Blackwell.

Smith, R. P. (1985), 'The significance of defence expenditure in US and UK national economies', *Built Environment*, 11(3), 163-70.

Snoddy, Raymond (1985), 'Executives neglect benefit of information technology', *Financial Times*, 17 September.

Soete, Luc and Dosi, Giovanni (1983), *Technology and Employment in the Electronics Industry*, London, Frances Pinter.

Storper, Michael and Walker, Richard A. (forthcoming), *Explaining Growth: Geographical Industrialization and Capitalist Development*, Oxford, Blackwell.

Takamiya, M. (1981), 'Japanese multinationals in Europe: internal operations and their public policy implications', *Columbia Journal of World Business*, summer.

Teulings, A. W. M. (1984), 'The internationalization squeeze: double capital movement and job transfer within Philips worldwide', *Environment and Planning A*, 16, 597–614.

Thompson, Paul and Bannon, Eddie 1985), *Working the System*, London, Pluto.

Tomlinson, Jim (1982), *Unequal Struggle: British Socialism and the Capitalist Enterprise*, London, Methuen.

Transnational Information Exchange (various dates), Amsterdam, Transnational Institute.

Turner, Louis (1982), 'Consumer electronics: the television case', in Louis Turner and N. McMullen (eds) *The NIC's: Trade and Adjustment*, London, Allen and Unwin.

UK Information Technology Organization (1986), *The UK Information Technology Organization*, London, UKITO.

Ungerer, Herbert (1986), 'The European Community's telecommunications policy: a global approach', in N. Garnham (ed.) *Telecommunications: National Policies in an International Context*, Windsor, Communications Policy Research Conference Proceedings.

United Nations Center on Transnational Corporations (1982), *Structure and Behavior of the Semiconductor Industry*, New York, UNCTC.

United Nations Center on Transnational Corporations (1983), *Transnational Corporations in the International Semiconductor Industry*, New York, UNCTC.

Urry, John (1981), *The Anatomy of Capitalist Societies*, London, Macmillan.

Urry, John (1986), 'Capitalist production, scientific management and the service class', in A. J. Scott and M. Storper (eds) *Production, Work, Territory*, London, Allen and Unwin, 41–66.

Vernon, Raymond (1966), 'International investment and international trade in the product cycle', *Quarterly Journal of Economics*, 80(2), 190–297.

von Hippel, E. (1977), 'The dominant role of the user in semiconductor and electronic subassembly process innovation', *IEEE Transactions on Engineering Management*, May, 60–71.

von Hippel, E. (1977), 'Industrial innovation by users', Working Paper 953–77, Sloan School of Management, Cambridge, MA.

Walby, Sylvia (1987), *Patriarchy at Work*, Cambridge, Polity Press.

Wales TUC (1984), *Evidence to House of Commons Committee on Welsh Affairs, The Impact of Regional Industrial Policy on Wales*, London, HMSO.

Walker, Richard A. (1985a), 'Class, division of labour and space', in Derek Gregory and John Urry (eds) *Social Relations and Spatial Structure*, London, Macmillan.

Walker, Richard A. (1985b), 'Technological determination and determinism: industrial growth and location', in M. Castells (ed.) *High Technology, Space and Society*, Beverley Hills, CA, Sage, pp. 226-64.

Weber, Alfred (1929), *Theory of the Location of Industry*, Chicago, Chicago University Press.

Willett, Susan (1986), 'The impact of defence procurement on the electronics sector in London', Mimeograph, Birkbeck College, University of London.

Williams, G. A. (1984), 'Women workers in Wales', *Welsh History Review*, 2, 530–48.

Williams, Karel and Williams, J. (1983), *Why are the British Bad at Manufacturing*, London, Routledge.

Wilmot, R. (1985), 'Wanted: an industry that is world class', *Financial Times*, 26 June.

Winckler, Victoria (1985), 'Tertiarisation and feminisation at the periphery: the case of Wales', Mimeograph, University of Wales Institute of Science and Technology.

Wong, Yut-Lin (1985), '"Oriental female", "nimble-fingered lassie", "women with patience": ghettoisation of women workers in the electronics industry', M.Phil. thesis, Institute of Development Studies, University of Sussex.

Index